The inside out of

flies

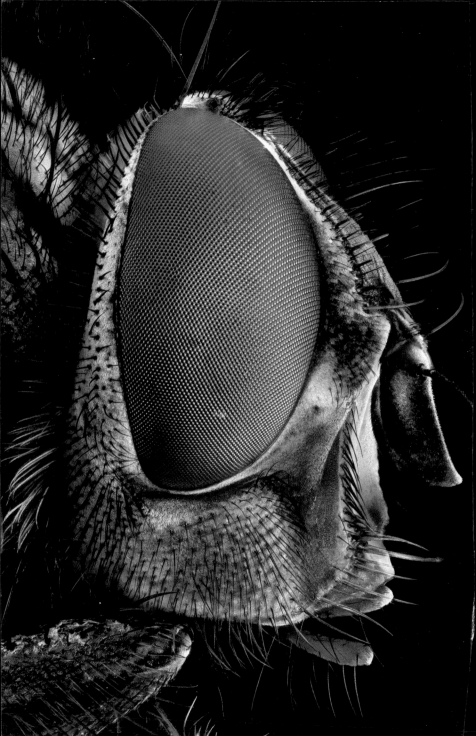

The inside out of
flies

Erica McAlister

Published by the Natural History Museum, London

I dedicate this book to Brian Stocks – I know that
his comments would have caused me to smile.

ABOUT THE AUTHOR

Entomologist Erica McAlister is a senior curator of diptera
at the Natural History Museum, London. Not just fascinated
by studying and naming the little things, Erica has long been
intrigued by how they *do* things, often inspired by the antics
that she has observed in the field, both on far flung tropical
islands and in her own 'island' garden in urban London.

Erica has studied in Dominica, Australia and Costa Rica and
her work with diptera has taken her all around the world. She
also presented the BBC Radio 4 series *Who's the Pest?* and
Metamorphosis – How Insects Transformed Our World.

First published by the Natural History Museum,
Cromwell Road, London SW7 5BD
© The Trustees of the Natural History Museum,
London, 2020.
Paperback edition 2021 with updates.

ISBN 978 0 565 09526 0

A catalogue record for this book is available from
the British Library.

10 9 8 7 6 5 4 3 2 1

Designed by Bobby Birchall, Bobby&Co
Reproduction by Saxon Digital Services
Printed by Toppan Leefung Printing Ltd.

Front cover: *Formosia moneta*, parasitic fly
Back cover: *Australoechus hellipr*, bee fly

Contents

Larva of M. DOMESTICA.

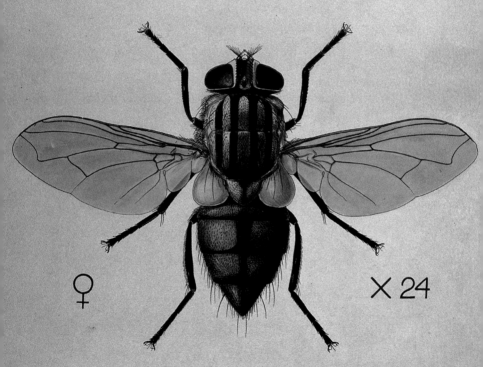

♀   × 24

A.J.E.
TERZI.

MUSCA DOMESTICA, L.

Introduction

Gretchen: You're weird.
Donnie: Sorry.
Gretchen: No, that was a compliment actually.

Donnie Darko, 2001

SITTING IN MY GARDEN on a wonderful August morning I noticed a large number of flies crawling around the ground in front of me. I picked one up gently and let it crawl and hop around my hand. It was a freshly emerged adult that had just burst out of its pupal case, its wings were still folded and its head was gently inflating. And I watched this fly for about twenty minutes. I watched as its inflated head sac collapsed and a face formed. I watched as the wings were stretched out and began to harden. And I watched it turn from a dull brown colour to a metallic green beast. I had watched this little beast transform in its final stage of metamorphosis from a wingless maggot to a flying machine, ready to make its inaugural flight in search of food and a mate. I had watched a blow fly become the male that he was destined to be. I felt privileged. In fact every time I look at a fly, I find something new to be fascinated about – their incredible diversity of form, their extraordinary range of behaviours and how they have set up home just about everywhere, including my garden.

Musca domestica described by Linnaeus in 1758 as having 'Os proboscide carnosa: labiisl lateralibus: Palpi nulli' (mouth with a fleshy proboscis: with two lateral lips: palpi absent).

I am not the only, nor the first person, to find flies stupendous. An ancient Sumerian poem, *The Return of Dumuzid*, talks of the Goddess Inana, who was helped to save her husband from demons by a fly. The Sumerians were the most ancient of the Mesopotamians, who lived in that region of the Middle East that includes modern-day Iraq, and their literature dates back 5,000 years. We have evidence that the appreciation of flies in ancient cultures was already prevalent during the New Kingdom (1550–712 BC), a hugely important period of Egyptian dynasties, featuring some of its most colourful of characters such as Tutankhamun, Ramesses II (The Great), Akhenaten and his wife Nefertiti. It's thought beautiful fly amulets were given to soldiers in honour of outstanding military service (though that's a matter for debate), as the fly was the hieroglyphic sign for determination.

This blossoming love for flies soon faded though, and throughout most of the remainder of history they have been associated with all things harmful, filthy, diseased and sinful – Steven Connor wrote in his popular book *Fly* that all they were seen as being good at was to 'steal, torment and seduce'. I strongly disagree with this statement - well, maybe not the last part – but still, I do think that they have had some incredibly poor PR, and this needs correcting.

I am fully aware that my opinion may not be the 'average', but who wants to be average? Flies certainly are not, and I hope to persuade you that this often-despised group of animals deserves way more kudos than they currently get. I hope to convince you that they deserve this kudos for the many fantastic ways that they have evolved to exploit not only their environment, but also other species and, more cheekily, their own. In understanding flies, we can understand so much more about our environment and many of the other species in it, including our own. Flies can get into every nook and cranny of life. They not only fly, but walk, swim, dive, jump, wriggle and dance. Some live just hours as adults, while others survive for more than 15 years as immatures. And to cope with such a diversity of traits they have developed a massive diversity of body forms, from those that are 0.4 mm in length to those that are nearly 8 cm (3¼ in) long. From the tip of their antennae to the end of their anal spiracles, flies are as fantastical as they are different.

This book aims to introduce the reader to the shape, structure and design of flies: from their eyes and the complexity of their vision, through the body, including the mechanisms of their flight, and of course their fantastic genitalia, a subject worthy of much discussion as we will soon learn. Just in terms of their shape or morphology, flies are some of the most varied of any insect group and contain some of the greatest body adornments an animal can possess.

In Book XVII of the *Iliad* Homer, the legendary poet from ancient Greece, describes the fly as daring, but mostly in a negative way, especially in reference to its blood-sucking habits. 'She (Athene)... gave him the persistent daring of a fly, that finds human blood so sweet that it keeps attacking however often it is flicked away'. Indeed, the majority of people on the planet today still think of flies as being associated with disease and filth. And I would be a liar if I said that this was never the case.

One family of flies in particular has a terrible reputation in terms of disease transmission. Culicidae is the scientific name for mosquitoes, and this family has been labelled as the deadliest animal on the planet. This is slightly unfair, however, as it is not the mosquitoes that are the killers, but what they carry. They are but pawns to the real nasties, the truly frightening bacteria, viruses and parasites. Mosquitoes have never directly killed a person. Only the females blood feed and on average she only takes five micro litres (that's about 0.001 teaspoons worth) during each feeding event. Say, for example, that the average human being, weighing between 70 and 80 kg (154–176 lbs), has between 4.7 and 5 5.1 of blood (8¼–9¾ pints). We can cope with losing some of that but once we pass 40 per cent things don't look so good. Undertaking further simple bar-mat statistics, it works out that you need about 440,000 mosquitoes to get to that stage of blood loss in humans. To date I have not read any papers where a human has suffered anything close to this degree. In fact, it would be physically impossible for this to happen during one feeding event as there would not be enough room on a human body for all the mosquitoes to feed. Death from exsanguination – the feeding of blood – has occurred in other animals such as sheep and cows; one of the worst mass killings by mosquitoes that I have read about ᴖᴖ in 1962, where the deaths of 1,550 cattle from Texas and Louisiana were attributed to the eastern saltmarsh mosquito, *Aedes solicitans*. But no perfectly drained human cadaver has to date been found. The mosquito has evolved into a highly effective stealth feeding machine and we can learn much from this. In understanding how it detects smells, feeds and reproduces we can not only apply this knowledge to controlling harmful populations, but we can further utilize this knowledge for many bio-inspired ideas, especially in the field of medicine.

But I need to cover some basics first, such as how we describe and name flies, and why these names are important. Humans first started studying and writing about flies about 2,000 years ago, when Pliny the Elder (AD 23–79) talked about them in what is regarded as one of the oldest natural history books ever written, his *Historia Naturalis*. In a beautiful turn of phrase Pliny writes that within the insects, of which the flies are a major component, 'no one of her works has Nature more fully displayed her exhaustless ingenuity'. But there are an awful lot of them and they are hugely varied, so folks set about giving names to all these organisms. One of the major hurdles to overcome was developing a universal system for naming them, a system whereby anyone across the globe could understand what animal or plant was being talked about, irrespective of their native tongue. It was only in the late 1700s that this was achieved, when the Swedish biologist Carl Linnaeus set about trying to order the life found on this planet.

Linnaeus was the first to stop using *Musca* as a general term for flies and to begin using the word as a formal name for a specific group or genus of animals, based initially on wing characteristics, though later he also considered mouthparts. He, like all good researchers, adapted and modified his names and classification system as he developed a greater understanding of his subject, and we have been adding and altering ever since then. Linnaeus is known as the Father of Modern Taxonomy, the science of naming, describing and grouping of organisms, and is arguably one of the most important scientific figures to have lived, thanks to such a simple system – well, certainly this *Homo sapiens*, and many others, think so. This universal classification system of ordering and naming, what is referred to as the taxonomic hierarchy, uses a binomial 'two names' naming system, where each living thing is given a genus and a species name e.g. *Homo* (genus) *sapiens* (species). These genus and species names were originally described in Latin as it was the language of science in western Europe at the time. Modern names derive from many different languages and so it's more appropriate to refer to this binomial name as their scientific rather than their Latin name. When we write scientific names in the literature, we don't just include the genus and species components, but often add the author and year that the name was published, as this helps us to ensure that we are talking about the correct species.

A species is often defined as a group of animals that can only breed amongst themselves, and a genus is a group within which similar species are clustered. I say 'often' as there are many examples where this is not the case, and I believe there are at least 26 different definitions of biological species. Species are the biological entities, the individual unit that we consider genetically distinct from their relatives. Anything above is what we perceive to be natural groupings (and different observers weigh different traits and morphologies differently), implying a hypothesized evolutionary relationship between the species into genera, into families, into orders, into phyla and so on. And we observers, we namers (taxonomists) and we researchers into how they relate to each other (phylogeneticists), are always changing this as new species are added into the mix and we develop new methods for re-analyzing old species (in the process giving everyone else headaches). Our scientific names reflect two things: firstly they help us know what we are looking at, help us to search the literature to understand this creature in front of us; and secondly, they reflect the current scientific hypothesis. This has not always worked in unison, and the classic example of this is with the most important research tool for biological research – the fly *Drosophila melanogaster*.

The ability of *D. melanogaster* to live and prosper in laboratory conditions as well as being relatively simple – it has just four pairs of chromosomes in comparison with our 23 chromosomes and just 132 million base pairs, the chemical cross-links between the two strands of DNA (deoxyribonucleic acid), compared with the 3,200 million base pairs we have – has made this species (and its siblings) an excellent research tool. As such it is referred to as a model organism, helping us to understand the molecular mechanisms of many human diseases as many of the basic biological, neurological and physiological processes are shared by humans and flies (75% of disease-causing genes found in humans also appearing in the flies' genetic code). Well, that and the fact it is very cheap to keep! Luckily for all, except maybe the flies, the combination of *D. melanogaster* reproducing quickly, being inexpensive to keep and being easy to maintain in colonies of large numbers, has over time proved its worth above all others again and again. Initially, research with *Drosophila* was dominated by questions associated with genetics, but with increased technological advances we have been able to look at other aspects such as nervous system development and function. By understanding the relatively simple systems of these flies we can begin to understand those of other flies and perhaps even those of our own. In her wonderful book *First in Fly,* published in 2018, Stephanie Mohr, of Harvard Medical School, writes of how researchers working

Drosophila melanogaster doing what they do best – providing hundreds of offspring to help us understand many aspects of biology.

on this species are so enamoured by their topic that they describe them as 'Intrepid. Loyal'. And the rewards are plentiful: just understanding their genes alone has brought major breakthroughs in our knowledge of inherited diseases, over 800 to date including Huntington disease, Downs syndrome and muscular dystrophy. I think that my favourite gene discovered by looking at *Drosophila* is the amusingly named cheapdate gene because it is particularly sensitive to alcohol (in both flies and humans). As of 2017 there had been eight Nobel prizes awarded to research on *D. melanogaster* and at this rate there are likely to be many more.

The *Drosophila* genus was described back in 1823 by the Swedish botanist and entomologist Carl Fallén. Fallén had taken a species that Johan Fabricius, a Danish entomologist and a student of Linnaeus, had originally called *Musca funebris* and created a new genus, *Drosophila*, using this species as the type species, i.e. the reference species. Seven years later, in 1830, Johann Meigen, a German entomologist and arguably the father of dipteran taxonomy, described a new species he was studying as *Drosophila melanogaster*.

Over time more and more species were added to this genus so that it has more than 1,450 species to date. But over time, there came a greater understanding that this genus had been used as a bucket – many species were thrown in but there was no structure, no understanding of how they were related to each other – and so these were divided into subgenera. These new subgenera included *Drosophila (Drosophila)*, where the original type species resides, and *Drosophila (Sophophora)*, where *D. (Sophophora) melanogaster* species resides (the etymology of *Sophophora* means 'carrier of wisdom'). The revision did not stop there. With more knowledge of this and other species, came the realisation that this genus is paraphyletic, a term that means that flies from different evolutionary backgrounds have been artificially lumped together. In correcting this situation, those subgenera that we knew were different would have to be promoted to their own genus which in this case would have resulted in the birth of the genus *Sophophora* and the species name *Sophophora melanogaster*. This is a complicated process and can result in some epic battles between researchers who disagree with each other and their proposals. In the business of taxonomy, we have what are referred to as lumpers or splitters, those that keep definitions broad and those that divide at the slightest change in hair length! And here it was a splitter that was spreading the seeds of change. But for this species, things were not so simple – folks argued that all of those millions of papers (ok nearly 50,000 as determined back in 2010), that had *Drosophila* in the title, would be hard to find if it was split, and all those research labs named *Drosophila* would be affected – many, many, many researchers working on this fly do not care about why it was named so, just that it was named and they wanted consistency with other workers. Oddly this argument has worked (at least for the time being) and indeed *Sophophora* has been maintained as a subgenus.

However confusing this scientific system of naming may seem, it is in my opinion, never as confusing as using common names. I know that many folks like common names and I'm always being asked for them for flies, but this is problematic for several reasons. Firstly, most simply don't have them. When you have a family of shiny black flies of say 300 species of very similar looking beasts you soon run out of suitable names. (shiny black fly, very shiny black fly, less shiny black fly, another black fly, a bold black fly etc etc). And although sometimes confusing, scientific names help people from around the world communicate with each other in one language. What's more, common names can suggest relationships that aren't there. Dragonflies, horse flies, mayflies, scorpion flies, bee flies (in the UK) etc may be easy names to remember, but only two are true flies. All insects, yes; flies, no. And to make matters worse, as a general rule of thumb those names that have the word fly integrated into them such as dragonflies and mayflies are the ones that aren't true flies. Horse flies, house flies, bee flies etc are the true flies, those that have just one pair of wings (diptera comes from the Greek 'di-' meaning two, and 'pteron' meaning wing). But this doesn't always hold true. Scorpion flies are not flies and many countries have reflected this by combining the words.

Even if the common name does refer to real flies, it may refer to a group rather than one individual. From autumn until early spring, we field enquiries at the Natural History Museum from concerned citizens about cluster flies – those slightly furry, yellowish creatures that aggregate in quite large numbers in our houses (can you believe the temerity of other species being inside our houses?). The problem with calling them by this name is that they are not one species. The term cluster flies refer to the genus *Pollenia*, of which there are eight species in the UK and thirty-one in Europe. And so we need these binomial names to help us determine exactly which species is which. Not all *Pollenia* have the same behaviour, indeed only those within the *Pollenia rudis* group cluster, and so in this case knowing the species is important, though not perhaps as important as with medically important vectors – the carriers of disease.

Looks alone are often not enough to divide species. The mosquito *Anopheles barbirostris* is found in the Oriental region, the name given to a specific biogeographic region, and is one of 13 species that look incredibly similar. We rely heavily on molecular techniques to examine species' genetic makeup and differentiate between them. And what these genetic differences are varies across organisms. There is no set number of mutations or changes that reveal the existence of different species; it depends instead on what the results of these changes are, e.g. extra bristles on their genitalia, and how they affect the behaviour of the species.

When I say that *Anopheles barbirostris* is a species, I am being slightly disingenuous, as it is part of a group of different mosquitoes that are so closely related that we find them hard to separate morphologically and sometimes genetically, so we have lumped them together as a species group or complex. This is a common problem

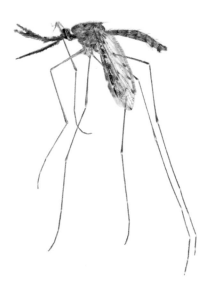

Anopheles barbirostris – one species or many species?

with mosquitoes and presumably with other, less studied groups of animals. This species and a further thirteen related species form the taxonomically complicated Barbirostris group of which all but six, including *An. barbirostris* species, are in a subgroup that are almost identical as adults but are thought to differ in their roles as malarial and filariasis vectors in Southeast Asia.

So it is important that we try and resolve these taxonomic headaches to help us effectively limit the spread of disease and in doing so we need to analyse the original type specimen of the subgroup, to ensure we assign the correct name to modern material. Samples kept in museums are sometimes decades old though, and with all due respect to my place of employment, they can be limited in what they are able to tell us. Thanks to wear and tear over time, many specimens in the collection have missing body parts which makes identifying any distinguishing features really hard. Fresh specimens serve us so much better when trying to resolve complex taxonomic questions, both in respect to their morphology and DNA. And so, several moons ago, when I was but a wee maggot at the Natural History Museum, London, I headed off to Java, Indonesia to hunt these babies down from the original locality that they had been collected from.

Armed with what looked like a Ghostbusters power pack (in fact a portable suction sampler that hoovers up the flies), I spent three frustrating weeks walking around the base of a volcano trying to catch them, a process that left me questioning my life choice (of being a fly botherer). One particularly low point was at 11pm one night, when I found myself vacuuming mosquitoes from a tethered but very amorous bull. It was half an hour of pure adrenaline as I tried to collect the mosquitoes from him without him gaining any access to me. I've never been more grateful for a sturdy rope! But through such exploits, including searches in cow sheds and sewage pits, paddy fields and stagnant pools, I eventually found the species I had come for.

My mosquitoes and I headed back to London (with our permission letters) where my esteemed colleague Dr Ralph Harbach, a world mosquito expert at the Natural History Museum, and Dr Harold Townson, Emeritus Professor at the Liverpool School of Tropical

Medicine, were not only able to give a comprehensive morphological description based on new specimens, but also a molecular one by determining their DNA profile, a process we call sequencing. And in doing so the original species, *An. barbirostris,* was divided into two, with the recognition that indeed within this species, there was both the original species and a new species, *Anopheles vanderwulpi* Townson & Harbach 2013. This complex species had just become more complex. But is this new species an important vector? Does it have the same behaviour as others in its group? We don't know the answers yet, but it looks increasingly like the original species is not as an important vector as we previously thought, but some of these new species, including *An. vanderwulpi,* may be the true villains. Now we know what species are found on Java we can start to work on these questions surrounding the distribution and vector capabilities of other members of the subgroup.

Morphology is very important and, when we describe a new species, we observe and list all the characteristics of its body to help us work out differences and similarities. Being able to describe a fly in detail, the precise way it looks from its head to the tiny hairs on its feet, is fundamental in ensuring that anyone else who finds one can correctly identify it. So many species descriptions cause heartache for researchers as they have been less than adequate in their descriptions.

At present about 1.5 million species of organisms on the planet have been described, a number that grows every minute with researchers collecting and identifying new ones. I am lucky to have spent time over the last couple of years on the beautiful island of Dominica in the Caribbean, collecting all manner of strange and shiny creatures. And in these collections, I know that I am probably looking at specimens down my microscope that belong to a species that has yet to be described, and so I too will be adding a few names to this ever-growing list.

The order Diptera, the true flies, is in the class Insecta, which forms part of the phylum Arthropoda. Ernst Haeckel, a German philosopher, biologist, physicist and naturalist, to name but a few of his talents, introduced the term phylum in 1866; groupings of genus and species were already around and based on what the organism looked like. At present, within the kingdom Animalia, there are 32 phyla (this

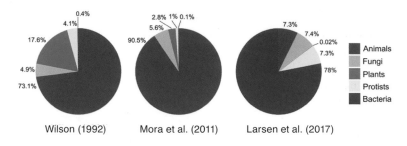

Species diversity of the main groups found on the Earth at present.

number varies though depending on classification systems, definitions and current thinking in science). The phylum Chordata contains us humans, along with all the other mammals, birds, fish, sea squirts, amphibians, reptiles and the fish-like lancelets. It sounds vast, but it is only approximately 65,000 species. Of the 1.5 million or so species that have so far been described on the planet, approximately 1.26 million of them are Arthropods. The Arthropod phylum includes the crabs, spiders, millipedes, centipedes, insects etc, and the combined species richness completely overshadows the relatively puny species richness of chordates! The true dominants will arguably turn out not even to be any animal but rather bacteria. Brendan Larsen, a virologist (studies viruses), who was part of a team at the University of Arizona, was lead author of a paper published in 2017 on the predicted numbers of Earth's species, which predicted that the number of bacteria will blow everything else out of the water and increase the overall number of species in the world to two billion. Dear Natural History Museum, we are going to need a bigger building!

Nevertheless, arthropods are incredibly successful. They dominate both the terrestrial and marine environments. They laugh at the fragility of birds, poke fun at the armour of mammals, and mock the lack of flexibility in snakes. Within the arthropods it is the insects that rule. All insects are divided into three main body sections: a head, thorax and abdomen. They differ from their closely related cousins, the spiders and the crustacea, by their number of legs and the arrangement of their body parts. Spiders for example, have fused their heads and thoraxes into one cephalothoracic (cephalo = head) unit, and have four pairs of legs. All adult insects have three pairs of legs (well at least to start off with – anyone who works with crane flies knows how easily six legs soon becomes less). But irrespective of leg numbers, they all share a feature that is one of the key reasons that these species have achieved such greatness and that is their body armament – the development of a chitinous exoskeleton or cuticle, which covers them, and both protects and supports them. The presence of this fibrous chitin makes for a much stronger exoskeleton, whether big or small.

One of the largest insects in the world is the giant weta (*Deinacrida heteracantha*), a type of cricket from New Zealand which can grow up to 7.5 cm (3 in) long and weigh up to 70 g (2½ oz), the equivalent of one and a half golf balls. The smallest insects are the fairyflies (again with the slightly unhelpful names as this is not a fly but a wasp), at just 0.139 mm (0.0055 in) long and weighing far less than an eyelash. The smallest fly is just 0.4 mm, so it is presumed a similar weight. Whatever the size of the fly, they have managed to adapt their shape to suit their needs in a multitude of habitats, and there is considerable variation in flies in terms of the rigidity and thickness of their cuticle across the different species, across the two sexes and during the different stages of their life cycle.

The exoskeleton of a blow fly maggot, the term for the juvenile stage is, for example, very thin and extremely flexible, enabling it to squeeze through the most minute of gaps. Poulomi Bhadra, a masters student working alongside Dr Martin Hall, a forensic entomologist at the Natural History Museum, was working on a project looking at how physical barriers can delay the arrival of maggots at a corpse. The barriers in question were zips on suitcases as, between 2001 and 2014 when her work was published, there had been at least 20 instances in the Greater London area where bodies or parts of bodies had been found in suitcases or holdalls (A. J. Hart, unpublished data 2014). What the police were interested in was whether or not these barriers would cause a delay or completely prevent the flies from finding what was in these suitcases, and thus affect predictions of time of death. To answer this question, she cut the zips off five different models of suitcase which varied in tooth size, composition and shape, and placed these zips over dishes of sealed meat (and some empty ones for controls). And then she released the adult, female fly. It seems the females weren't deterred by the zips at all, as they either inserted their egg-laying tube through the zips and squirted their eggs directly onto the meat or laid them nearby. These zips provided no real barrier to the larvae either as they squeezed themselves through, their natural plasticity winning the day. This plasticity is a very useful adaptation for an individual to have as it enables it to find and exploit new resources. The adult stage of a fly is more rigid, with its external skeleton, the cuticle, being divided up into plates. The cuticle, though stiffer than a maggot's, may be a lot thinner in comparison to a lot of other insects. This allows for a greater flexibility in movement, a useful trait that has led to these creatures becoming the most agile of fliers.

One of the many amazing aspects of flies is that they rarely follow the rule book, and many, many, many species look nothing like a typical house fly, the species most people envision. The insects with the greatest variability in their diets seem to have also some of the greatest variability in their shape. Take those basic features that we use to identify a creature as a fly and it all becomes messy; some adult flies have no

Going somewhere? Maggots growing through zips in a project looking at how and whether physical barriers can delay or prevent flies finding a corpse.

mouthparts, many species lack wings or just have simple protrusions and some lack both wings and halteres (their balancing organs that I talk about in Chapter 6).

The range of eye modifications alone is staggering; eyes on the end of stalks evolved independently a whopping 22 times. But the parts that fascinate me the most are the genitals. I mean, there are species of fly whose males appear to have genital whips and others with big feather duster-looking structures. And not to be outdone by the males, females have impressive egg-laying structures, called ovipositors, some of which resemble tin openers. The ability of the flies to evolve quickly and modify their genitals has let this crazy bunch of species keep their populations booming.

All these modifications and more have enabled flies to become one of the most species-rich and ecologically diverse groups of any animal on the planet. They are one of the mega rich orders with, to date, 160,000 or so species described from approximately 190 families. No one can quite agree on the number of families, but whatever it currently stands at, we can be sure this current species estimate is woefully below the real figure. For example, a study completed in Canada in 2016 by a team led by Professor Paul Hebert from the University of Guelph, wanted to determine a good estimate of species richness for Canadian insects. Over a period of 10 years they collected and identified, from all across Canada, several million specimens that were sorted and identified to the highest taxonomic level possible and barcoded. They found that both Hymenoptera (bees, wasps and ants) and diptera were far more diverse than previously suspected, nearly doubling estimates from 54,000 species of insect in Canada to 94,000. But within diptera, there were further surprises for the researchers. The family Cecidomyiidae or gall midges, previously thought to have 6,600 described species worldwide, turned out to have 16,000 in Canada alone.

Now there are many reasons to be cautious with all these types of species estimations, but the take home message is that there are a lot more species than we think. The challenge is to identify them when we find them. For instance, many flies in the previously mentioned gall midge family are nigh on impossible to tell apart if you only have females.

Now, I am probably going to be disowned by the die-hard dipterists for disclosing secrets, but I have been on many a field trip where the average fly collector has bemoaned the presence of females in their sample as their genitalia are perceived to be so similar as to make identifications for us mere mortals virtually impossible. The male genitalia, on the other hand, comprise many more complicated parts in comparison to the female.

Gall midges are not unusual; the identification of females as well as the immature stages, the egg and pupal stage as well as the larvae, have all been neglected. The advent of new imaging techniques for looking at the specimens, and the rise in molecular methods as well as other less morphological approaches will hopefully start to fill in these gaps. The lack of identification of the immature stages is very problematic, as this is often the longest part of a fly's life and often the part that has the greatest interaction with its environment as it is its major feeding stage. Malcolm Butler, an American researcher on midges, wrote about species in the genus *Chironomus* that live in the harsh tundra pools of Alaska and take seven years to reach adulthood. Among the most inventive ways flies use to survive challenging weather is to stop developing

An example of the variety and form of galls caused by developing midges from the family Cecidomyiidae.

altogether, to enter a period of arrested development called hypobiosis. Another chironomid, *Polypedilum vanderplanki*, takes a leisurely 17 years to develop, albeit spending most of the time in a desiccated state! Once more the incredible adaptability of flies seems almost unreal.

The separation of life stages has benefited the insects, especially flies, enormously and has led to rapid and major speciation. Mammals, with around 6,500 known species, often have to compete with their offspring for resources. Flies mostly don't have this problem. The larval stages of most flies have a completely different diet to that of the adult and occupy different niches, so further reducing competition for resources.

Take the wonderful bee flies (Bombyliidae) – these truly are the poster species for flies, resembling a flying powder-puff. The adults are varied in form, but if they feed (some don't) they feed on nectar and sometimes pollen using their long straw-like mouthparts, or proboscis. But their larval stages (there are more than one) are not so vegetarian in their diet and target many species of invertebrate including beetles, spiders, wasps and, much to the public's horror, bees. They are parasitoids, creatures whose feeding behaviour always results in the death of their host. To enable them to feast on such creatures, who are often much more robust and armoured, they have retained their primitive mouthparts, the mandibles, but these now resemble lance-like hooks with teeth carved into them – brutal apparatus perfectly suited for penetrating the cuticle of their hosts.

Clever modifications across the species and families of flies have enabled them to adapt to many different, bizarre and extreme environments. Flies are able to cope with flying at extreme speeds through these environments, processing information at a rapid rate, and of course my favourite, being able to flirt in even the most challenging of conditions, often at a fast tempo. Inside and out, they are finely tuned to living well and breeding fast. This book will take you through the bodies of these little wonders, from their antennae to their anal spiracles, from the commonplace to the extraordinary, showing you how modifications have enabled the proliferation of these species, and also inspired us humans in the development of our own technologies.

CHAPTER 1

Pre-adulthood

Summer will end soon enough, and childhood as well.

George R R Martin, A Game of Thrones

T O BECOME AN ADULT – that fun stage of life where they get to travel around and fornicate – involves overcoming a few obstacles. The first, and perhaps most crucial, is to make a success of being immature. Unlike Peter Pan, the aim of most species is to reach adulthood, and to do that they need to lay down enough reserves for the transformation from immature to adult to take place.

Every maggot begins life as an egg, the part of the life cycle where they are least able to defend themselves against predation and the elements. Immobile, and often very small, eggs must survive desiccation, freezing, drowning and countless more environmental anxieties. The majority of fly eggs are tiny, generally one millimetre or less – smaller than many other insect eggs, and generally unseen by most people (the obvious exception is the eggs of those flies that are attracted to rotting or exposed meat and cat owners know these eggs all too well). The smallest are what we refer to as microtype, the tiny variety found in some of the genera of the family Tachinidae (tachinids or parasitic flies). At a microscopic 0.02 mm (0.00079 in) in length, they are a fraction of the size of the full stop that comes after this sentence. The biggest known eggs are found in the flesh fly family, Sarcophagidae, which although closely

Atherix ibis' hairy projections look remarkably like a Donnie Darko hallucination.

related to the tachinids, produce eggs that can reach 2 mm (0.04 in) long, a massive several full stops put together! You won't see the eggs of this family though, as they hatch inside the mother. She is what is termed ovoviviparous (or sometimes larviparous), where the eggs develop within the body and either hatch internally, with the larva being nurtured until they are about to pupate or hatch as soon as they are laid.

Many species of fly that are ovoviviparous have fewer eggs than those that lay their eggs externally, as the former takes a lot more time and energy per juvenile. Take the group Hippoboscidae, a wonderfully alien looking family of flattened flies that live and feed on the blood of birds, bats and large mammals, which give birth to live young. One of its species, the louse fly, *Hippobosca variegata*, produces an average of just 4.5 offspring during its lifetime, although other family members do produce more. This is a very different figure to the hundred or so that some non-ovoviviparous groups lay, where very little input is made by the parents after the eggs have been laid. In 1967, Robert MacArthur

The eggs of a bluebottle (Calliphoridae). A female may be more than half a kilometre away and still find food for her offspring.

and Edward O. Wilson, two North American ecologists (and more), wrote about this phenomenon, which they called the r/K selection theory. The decision to make quality over quantity results in the selection of different traits: for K-strategists such as the hippoboscids, there is the expense of greater individual parental investment to the adult female in terms of nutrient requirements and so on, whereas for the r-strategists, e.g. the bluebottle flies discussed and also groups like Tabanidae or the horse flies, there is a reduced quality i.e. very little nutrient and protection for the larval stage, but the adults can produce large numbers to compensate. r-selected species are seen as more opportunistic and are able to propogate quickly if the environment is suitable, whereas the K-strategists spend more time investing in a few offspring to give them whatever competitive advantage they can to prosper in what are often already crowded niches. Which strategy is chosen is very dependent on the environment, and in some cases can be adapted if the environment changes.

The many perfectly arrange eggs, glued to a stem, of an r-strategist species within the family Tabanidae (horse flies).

From each egg that successfully hatches comes a larva, also known as a maggot. This, more often than not, is the longest period of a fly's life cycle and is driven by just two things: for the larva to eat as much as it can and do so without getting eaten itself. So its body is nothing more than a basic eating machine, with no wings, no genitalia and no true legs. These structures are necessary during the adult stage, to aid with migration, dispersal and reproduction, but are not needed for the larva. Larvae live in a huge variety of habitats, you name it and they probably live in it. From the silt at the bottom of lakes to leaves in the tallest canopies, from a decomposing cow to a camel's nostril, from a spider's abdomen to a toadstool that the host sits upon, larvae seem to get everywhere. To cope with such a massive variety in habitat there is a huge variety of different forms, so much so that you can divide these mini feeding-machines into groups depending on what they look like, specifically by the shape of the unlike adult flies which can be divided into two major taxonomic divisions by the structure of the adult antenna.

Flies started appearing around 250 million years ago. They were originally split into two suborders reflecting their appearance (both in terms of how they look and their evolutionary timescale) – the Nematocera and the Brachycera. We know now that the Nematocerans are not related to one common ancestor and so we loosely refer to this collection of infraorders as nematocerous or lower flies. But, as Brian Wiegmann, one of the many research scientists working on trying to resolve these relationship issues (he is the 'Agony Uncle' for flies), points out these terms are fuzzy, but for ease in this book, I will refer to this old suborder as the nematocerous flies. The larva of many of these species have a distinct head casing, and most resemble leg-less caterpillars or nematocerans. Within the Brachycerans, these cases became smaller, and less distinct or complete (we refer to these larvae as hemicephalic), until the head case completely disappears altogether (the acephalic larvae). For example, with the house flies and their relatives, there is no distinct head capsule; instead, there is what appears to be just a pair of pirate's hooks that they use for a mouthpart.

A larva's life is to eat and grow, and to grow it must undergo a series of moults, to create room for all the internal development and the laying

SUBORDER	CURRENT GROUPINGS	LARVAL HEAD DEVELOPMENT
Nematocera (lower Diptera)	Nymphomyiidae[3]	
	Deuterophlebiidae[3]	
	Tipulomorpha[1]	
	Ptychopteridae[3]	
	Psychodomorpha[1]	
	Culicomorpha[1]	
	Perissommatidae[3]	
	Bibionomorpha[1]	
Brachycera (higher Diptera)	Xylophagomorpha[1]	
	Tabanomorpha[1]	
	Acroceridae[3]	
	Hilarimorphidae[3]	
	Stratiomyomorpha[1]	
	Asiloidea[2]	
	Empidiodea[2]	
	Apystomyiidae[3]	
(Cyclorrhapha)	Muscomorpha[1]	

The original grouping of flies into the two 'suborders' alongside the current groupings including the infraorders[1], superfamilies[2], and families[3] (adapted from Tree of Life and Wiegmann and Yeates (2017)).

down of fat reserves. The periods between moults are called instars, and the exact number of these vary from species to species. During each moult, the larva sheds it skin (and within the nematocerous species, the external head capsule is shed, too). The shed skins eventually decay but, interestingly – and as it transpires helpfully for us scientists – the head capsules endure. In fact, they can remain intact for millions of years as they are made of sclerotized chitin which is very durable and, in the case of one specific fly family, the Chironomidae, or non-biting midges, have helped scientists to understand previous climatic conditions.

I first studied the Chironomidae as part of my doctorate research on the ecological entomology of wetlands, and it is because of these species that I found myself working at the Natural History Museum today, thanks to a Dr Steve Brooks. To gain help in identifying them, I turned to Brooks, then a researcher at the museum, and one of the world's authorities on this large family of over 10,000 species (the same number as all the described birds on the planet), for help. Their larvae are incredibly important in freshwater ecosystems, both because they provide food for so many other species (my then head of department used to refer to my PhD as just research into bird food), but also due to their own feeding habits. These little marvels are some of the best shredders of fallen wood, some of the most active munchers of algae, some of most dominant feeders of detritus (as well as being kick-ass predators) and so essential little nutrient recyclers or waste-removal units. But often these environments can be very low in oxygen because, thanks to the high nutrient levels in the water, those ever-so-efficient algae and bacteria have been on a feeding frenzy and already used a lot of it up.

Haemoglobin, the oxygen carrier in animals, is not only found in vertebrates – creatures with backbones – but also sporadically throughout the invertebrates, including the nematodes, molluscs and crustaceans and insects (the other groups use different oxygen-carrying materials). Barbara Walshe, a researcher at Bedford College, London, was an early pioneer into understanding Chironomidae haemoglobins in the late 1940s and early 1950s. She memorably wrote that the haemoglobin proteins seemed to have been 'spasmodically distributed among the invertebrate phyla', i.e. it appears in some species, families and phyla but not others. In chironomids, and other arthropods, the main component is erythrocruorin, a relatively large (3.5 million daltons or 5.81^{-15}mg) protein. Generally, in vertebrates, this oxygen-carrying protein is housed in specialized cells called erythrocytes (it's what gives the red blood cells their colour). But in invertebrates not all the protein-carriers are confined in cells, with many roaming free, and there's a whole lot more variety going on with shape and size. These protein carriers can be quite big in chironomids in comparison to those found in vertebrates, as well as quite varied – *Chironomus riparius* has eleven types, we humans have one. It's not just their appearance that varies but also their function – these molecules do not just transport oxygen, they also store it, enabling these wonderkids to survive in extremes of different oxygen levels.

Humans and the other vertebrates breathe by using lungs to power oxygen into their bodies and then a transport system carries this, via the blood stream, around their bodies. Some arthropods also have a lung system, albeit very primitive, such as the box lung of spiders, or the feathered gills of crustaceans, which draw in large amounts of air. Some larvae also have small gills (e.g. *Atherix ibis* [Athericidae]), while others absorb oxygen directly through their cuticle. But most flies, and other insects, absorb oxygen through holes called spiracles that run down the side of their bodies. The number of holes depends on the species and the stage of lifecycle. The oxygen then flows through body through a network of tubes referred to as the tracheal system, enabling these essential molecules to get to all parts of the insect. The number and position of spiracles varies across the orders, for example the immature chironomids have none at all. This is not a common set-up though: two or eight pairs of spiracles are more frequent. Chironomid larvae are termed apneustic (from the Greek 'a' meaning not and 'pneusis' meaning breathing). Although

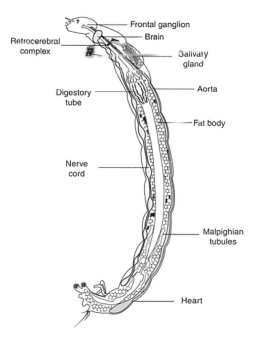

The internal morphology of the larva of *Chironomus sancticaroli*.

Labels: Frontal ganglion; Brain; Retrocerebral complex; Salivary gland; Digestory tube; Aorta; Fat body; Nerve cord; Malpighian tubules; Heart

they have no spiracles, they still have the tracheal system into which the gaseous oxygen diffuses into the cuticle.

Insects, unlike us and our neatly tubed (closed) circulatory system, have a much more unstructured system called an open circulatory system. For much of the time the insects' blood, the hemolymph (containing all manner of salts, nutrients, hormones and the aforementioned haemoglobins), is free flowing around its insides where it directly flows into the organs and tissues. There is a dorsal vessel that runs through the body, which collects the hemolymph in the abdomen and moves it forward to the head. As such, it is the abdomen that pumps – yep, flies have their hearts in their bums! For most of the time this system is adequate for survival (there are further pumps for structures such as wings that are discussed elswhere). However, species that live

One of the blood worms, *Chironomus riparius*. Its rich colour is due to the high levels of haemoglobin it contains.

in oxygen-poor environments, such as the chironomids, have to rely on additional support from their respiratory proteins, which carry more oxygen to different parts of the body and store it for when it is needed, and so different species are found in different oxygen environments.

We can identify the chironomids to species based on their head capsules and, after a lot of study, researchers have been able to determine which species are able to survive in environments with different oxygen concentrations, thanks to their haemoglobins. And not just survive but prosper – the densities that some of these species reach in some pretty extreme environments is quite something. For example, a 2011 paper published by Brett Ingram, a researcher at the Victorian Fisheries Authority near Melbourne, records chironomid larval densities of more than 73,000 per square metre. That's a huge amount, but figures nearer 188,000 individuals per square metre were recorded in the colder climes of Lake Myvatn in the Icelandic countryside in 1973 by Claus Lindegaard and Pétur M. Jónasson, chironomid experts from Denmark and Iceland respectively. It's not just clean, cold lakes they like, but also dank, oxygen-poor environments such as agricultural, domestic and industrial effluent – the tastiest of habitats. Different species of midge have adapted over millions of years to living in differing environmental concentrations of oxygen, and by knowing the oxygen requirements of certain species today, we can determine what the concentrations of oxygen were at specific times in the past. This is what Steve Brooks terms a 'midge thermometer', to help us understand past climatic events and previous atmospheric concentrations of oxygen.

Being able to breathe, either in freshwater, the sea or on land, is a fundamental process and, I think that the adaptation in spiracles or breathing holes across the flies is truly extraordinary. Take a moment to think about the larvae that live in another environment, that of inside other creatures, be they plants or animals. We are not just talking a few species of flies, but thousands of species. Within parasitic flies, spiracles that enabled the larvae to breath internally developed independently in 21 families, but happened just once with the parasitic wasps.

To breathe, parasitic larvae have adopted two different strategies. The first is to form a robust tube between the surface of the obliging host that extends deep into the body, allowing the larvae to breathe the further they penetrate, and a fun example of this is with another tachinid, *Exorista larvarum*. This species attacks the caterpillars of many different kinds of moth, including *Cydalima perspectalis,* the box tree moth. This moth has recently started turning up in the UK and is an invasive species from Eastern Asia that feeds on box trees. I have a massive soft spot for this tree and so I welcome this little maggot crusader in taking down this pest species. The first and second instar larval stages of this fly cause an immunological response in the host, tricking it into building a breathing passage, to which the larva attaches itself using some anal hooks. By the third instar stage, it abandons this passage and moves freely around its now dead host.

The second strategy adopted by parasitic larvae is to breathe directly by 'hacking' into the host's breathing system, a technique used by other Tachinidae. One species in particular, *Compsilura concinnata*, has adapted ways of penetrating more than 200 species of beetle, bee, wasp, moth and so on while using the host's tracheal system to directly steal oxygen. Ryoko Ichiki and Hiroshi Shima, two Japanese entomologists, published a paper in 2003 on their observations of this species doing just that to a silk worm (*Bombyx mori*) caterpillar.

Larval spiracles can be highly modified and come to resemble anything from hairy carrots to fighter pilots in the First World War (well their anal spiracles do in any case!). In my first book on flies, *The Secret Life of Flies*, I wrote about the spiracles of the larvae of Tipulidae, which looked like Tamagotchi characters. Tipulids are not alone in having such strange adaptations, all of which aid in respiration in often very oxygen-poor environments. And many larvae increase the size of their spiracular slits as they develop.

In the flesh flies, there is the species *Sarcophaga (Liopygia) argyrostoma* which, like many of its relatives, develops in dead and decaying flesh, often in large numbers, all wriggling around competing for the most nutritious spot. The first instar larvae have just two slits for each of their terminal or anal spiracles, and for the 20 hours or so that this stage lasts, that is good enough. Once onto the second instar stage, though, they need a greater supply of oxygen to supply their rapidly growing bodies and this leads to more spiracular slits developing. The anal spiracles also slightly thicken, even more so in the third instar with a further spiracular slit forming, so resembling an oddly coloured plate of toad-in-the hole. (For all non-English readers that is a delicious dish of sausages cushioned in a giant Yorkshire pudding, which I admit, may not be appetizing in this context!) The crucial point of information for scientists is that these spiracles are morphologically different across species and so can be used for identification purposes, especially important in forensically or medically important species. And these spiracles are not always just slits. As with the tipulids, they can be more obvious protrusions that maximize the amount of oxygen being absorbed from the environment.

Although often amusing in appearance, these long breathing protrusions are essential for many species that live in aquatic regions, enabling them to breathe surface rather than diffused oxygen. The wonderfully named rat-tailed maggots are found in the hover fly

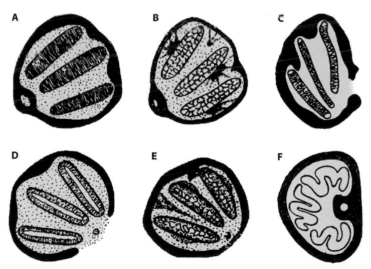

No, not the footprints of ducks but rather the posterior spiracles of the third-instar larvae of myiasis-causing flies: (A) *Calliphora* sp., (B) *Lucilia* sp., (C) *Sarcophaga* sp., (D) *Phormia* sp., (E) *Cochliomyia hominivorax* and (F) *Musca* sp.

family, Syrphidae including the genus *Eristalis*, and these are often seen squirming in our compost. These have long anal protrusions to enable them to breathe while submerged in this waste. These little creatures, thanks to their fondness for faeces, earned themselves the nickname of 's***apillars' by revellers at Glastonbury Festival in 2017. But toilet gags aside, their protrusions may help us.

These hover fly larvae have an incredible ability to decompose our waste and that of other animals, but while they live in fetid environments, incredibly, they don't appear to pick up bacterial infections. Like other plants and animals, this ability to resist surface infections is because they are covered in tiny hairs. While the hairs on species such as cicadas repel water to keep bacteria at bay, described as superhydrophobic, the hairs that cover rat-tailed maggots are hydrophilic, so they attract water, using it to form a barrier between the maggot and the damaging bacteria. A 2016 paper produced by Matthew Hayes, a PhD student at UCL and co-authors, published the results of their study that used electron microscopes to look at the surfaces of many things, including rat-tailed maggot hairs, known as nanopillars. They found that the nanopillars protected the maggot 'by offering a physical barrier between micro-organisms and the cuticle, in much the same way that anti-pigeon spikes prevent pigeons nesting on buildings'. These nanopillars stretch and deform the outer membrane or biofilm of the bacteria, causing it to rupture. Artificial barriers such as these are now being tested (although not yet with the maggots). Currently 700,000 people a year die globally due to bacteria-resistant infections, a figure that is predicated to rise to 10 million by 2050 if we don't find some way to combat the assault (and antibiotic resistance). Maybe, one day, those wonderful s***apillars will help save lives in all sorts of applications including antibacterial materials.

The head capsules of different species of maggot can be very varied too, and the Chaoboridae, the phantom midges, have particularly interesting ones. This small family of flies has a global distribution, but only 50 or so known species. They are closely related to chironomids and, although the adults do indeed look very similar, the larvae certainly don't. Instead they look like little humanoids in straitjackets as they swim about in water, and have the common name glassworms because

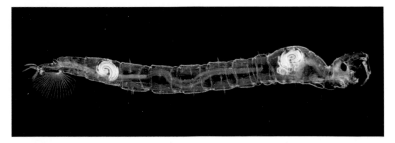

The phantom midge larva, *Chaoborus flavicans*, with its not so phantom prehensile antennae.

they are transparent. These species can grow up to 2 cm (¾ in) long and are generally predacious, although some are omnivores, all feeding in sediments on the bed or while free swimming. They may even do both, which is an unusual trait for aquatic larvae.

Irrespective of food preferences, all species in this family have head capsules from which an amazing pair of antennae arise and droop down in front of their faces. What is amazing is not so much their appearance but that the antennae have been modified to enable the larvae to grab prey. Having prehensile appendages like these, ones that can grasp, is not unique to this family of flies, but it is not that common among animals. The larvae swim along and strike out at prey, including other water organisms with prehensile antennae, for example the water fleas *Daphnia* spp., whereupon an epic, often fatal, tug of war on a tiny scale ensues.

Another family, the Simuliidae or black flies, also have aquatic larvae, but, unlike the previous examples, these are mostly found in fast-flowing rivers. The black fly family contains more than 2,200 known species globally, of which 1,800 are described from one genus, *Simulium*. We encounter the adults more commonly than the larvae as the adult females are all blood feeders, and as such many are nuisance biters or are complicit in the spread of diseases such as river blindness. As with mosquitoes and malaria, they are once more just a vessel used to transport the parasitic worm, *Onchocerca volvulus*, to an unsuspecting human.

If time is taken to observe black fly larvae though, a wonder can be seen – many of them resemble handle-barred moustached, spaghetti-

The labral fans of the head of a black fly larvae.

faced beasts when seen from the side, or pompom-headed cheerleaders when viewed from above, due to the presence of four fans located on the head. The great man of Simuliidae studies was Roger Crosskey (1930–2017), a researcher at the Natural History Museum, London, and a world authority on this family. In his wonderful book *The Natural History of Blackflies* he described these fans as 'large, obvious, intricate structures of considerable beauty' – and I for one, am not going to disagree with him.

Crosskey also writes that the larvae had mostly been described as sedentary in previous works, but in fact this is often far from the truth. These larvae can be found in a number of habitats – attached to stones on the bottom of rivers, or more amusingly, on the backs of other aquatic animals, anchored by a disc at the end of their abdomen that is lined with hooks. These phoretic (from the Greek 'phoras' meaning bearing and 'phor' meaning thief), larvae resemble members of a rather slow-paced Hells Angels gang, with their fans splayed out in the flowing current as they ride, not on a motorbike, but on a crab or a mayfly, to catch food (the hosts feed also on organic matter and are useful to the larvae because they are so much faster at getting to new sources). When they are not looking like little bikers, they appear to behave like cowgirls and cowboys, as they can also produce silk with which they can lasso around an object to use it as an anchor while they dangle, or drift along tethered by a thread.

The majority of black fly larvae are suspension feeders, and this is where those fans come in. They help the larvae to catch food from the water, along with a proleg – a single stump-like structure on their underside just below the head capsule – that helps concentrate the flow of food particles over the fans. The fans are covered in a sticky substance that has been secreted by the cibarial glands, so named as they are found in the cibarium or mouth cavity, the space just before the true mouth. The fans are regularly cleaned of food particles by mandibular brushes and hairs on the lip or labrum, all the caught food then passing into the digestive tract.

There is no clear relationship between head type in larvae and habitat requirements. Most nematocerous larvae have a head capsule and are predominantly aquatic, whereas the Brachyceran larvae have more terrestrial habitats. There are many examples where the opposite occurs even when the rest of the family has terrestrial larva. For example nematocerous flies such as crane flies and fungus gnats having terrestrial larvae. But even if they are in terrestrial environments, they all need moist habitats as they are prone to desiccation: a fresh dung pat, a rotting apple, in the leaf litter or soil are all terrestrial but are moist enough to ensure the little ones survive.

A good example of a semi-aquatic hodgepodge is the Tabanidae family, the horse flies. Species in this family live in a diverse range of semi-aquatic and terrestrial habitats including muddy banks, in the sands on the sea shore, and in gunk-filled tree-holes. The first two larval instars don't feed, but come the third instar – boy, do they become active

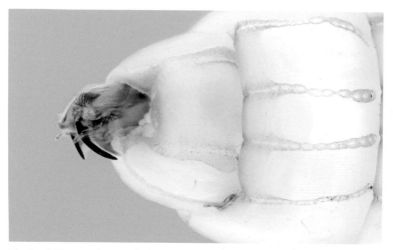

The hook-like mouthparts of a horse fly larva – they not only sink into their prey but also inject a paralyzing venom.

gluttons. By then, the majority are aggressive, active hunters, sporting small but brutal mouthparts. The diet of most horse fly larvae consists of a variety of invertebrates, but at least one species has larger prey in their sights. The larvae of *Tabanus punctifer* are not to be toyed with – there are accounts of even humans being 'stung' and they have been described as 'extremely aggressive and when handled carelessly, they can draw blood' – consider yourself warned. I love the fact that a creature smaller than a toe can produce fear and trepidation in an adult human. All horse fly larvae have vertical rather than horizontal mandibles, and they point outwards from the head, similar in appearance to the mouthparts of the mosquitoes and midges and look like a hook or sickle.

Occurring throughout North America, *Tabanus punctifer* larvae have been shown to predate on the New Mexico spadefoot toad, *Scaphiopus multiplicatus*. And I am not talking tiny tadpoles here, but newly metamorphosed adults, that range in size from 3.8 to 5 cm (1½ to 2½ in) in length. Roger Jackman, an English natural historian, while out collecting for beetles one night with several entomologists from Cornell University, observed many individual toads that appeared to have been pulled back into the mud, with at times only their head sticking out. Incredibly he goes on to describe that in trying to remove some of the 'fresh carcasses' out of the ground with forceps, they always met resistance – the tiny tenacious larvae were holding on to their prey from the other side and were not giving it up without a fight!

Interestingly these researchers weren't looking for frogs or flies but rather bombardier beetles – formidable creatures in their own right that squirt out an extremely volatile mix of hydroquinone and hydrogen peroxide at prey, which produces a substance that can reach close to boiling point. Not that the tabanid youngsters are put off – they also feed on these beetles. Apparently assisted by external head brushes, the tabanid larvae bite on to their prey, dragging them down, while pumping them with venom, partially paralyzing and quickly disabling them. Then they dissolve their insides. Yes, many dipterans are venomous. Researchers have known since the 1930s that tabanid larvae are venomous, but 90 years later we still understand little about these venoms.

The larval stage, though not as impressive as the highly motile adult stage, can still get around. They don't have 'true' (jointed) legs (Arthropoda comes from the Greek 'arthron' meaning joint and 'pous/pod' meaning foot) but many have prolegs as well as other adaptions which enable them both to move and to anchor to a spot. Prolegs are found on many immature insects, just think of caterpillars. They aren't 'real' legs; they have no articulated joints and have none of the muscles and nerves that the adult legs comprise, instead they are moved by flexion in the body. These legs may have spines, hooks or suction discs and often are used for both locomotor and sensory functions – those are not spiky baubles stuck on the larva, but the end of its prolegs. Other dipteran larvae have suction discs to enable them to stick around, essential when in a fast-flowing stream.

Blephariceridae, commonly called the net-winged midges, are a family of around 200 species whose larvae live in very fast-flowing streams – of up to 2 m/s – and so have had to develop many specialized organs to enable them to hold on. Not content with one sucker to do the job, they have six. And some larvae have something called a 'creeping welt' in place of a proleg. This is a sort of raised pad with spines. Its amusing name sounds more like an invasive human pox rather than

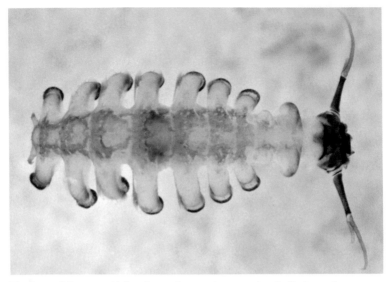

The larva of *Deuterophlebia shasta* showing its exceptionally fleshy prolegs.

what are basically suction cups that enable the larvae to move along the substrates surface. Although all these prolegs may look odd, the family with perhaps the oddest-looking prolegs is the Deuterophlebiidae, or the mountain midges (the first species, *Deuterophlebia mirabilis*, was found at an altitude of 3,566 m (11,700 ft) in Lake Gungobal, Kashmir, India). This family contains maybe 15 species, all in the same genus *Deuterophlebia,* and live across central and eastern Asia, and western North America. They are truly an enigmatic family, often living in steep and torrential mountain streams. All along the abdomen of these larvae are very large prolegs that can be turned inside out, and which protrude out of the side, much like a rib cage. And at the end of these prolegs are rows of between seven and thirteen long, flattened, small curved hooks, called crochets, that extend in and out.

These creatures walk along in a characteristic zig-zag pattern caused by alternately relaxing and contracting their posterior and anterior ends (and, if needs must, they can do this quite quickly). These enigmatic little creatures spend most of their short lives walking and grazing, pausing once in a while if the food source is rich. Not much is known about them at all (something uttered often by many a Dipterist), Harry D. Kennedy, an American entomologist working for Convict Creek Experiment station in California, wrote about their biology and behaviour in 1958. He found that most individuals had short lives, with three generations occurring across the year, and the first and second generation only living for 19 days. The third overwintered, but he was unsure during what part in their life cycle this occurs.

Chironomid larvae use their prolegs not to walk, but to swim, as do many other aquatic larvae. They swim using a side-to-side flexure, and this movement can also be seen in mosquitoes and a few other families. The biting midges in the family Ceratopogonidae resemble tiny bits of wiggling spaghetti, as they swim in a sinusoidal – an undulating s-shape – fashion. Some of the funnier ways terrestrial larvae get around, from my point of view, are found in some cecidomyiids, piophilids (the cheese skippers) and Ceratitis capitata (the Mediterranean fruit-fly, Tephritidae): the larvae jump! The larva loops itself round by contracting the ventral muscles that run from its head to its abdomen, and then attaches its mouth hooks to the cuticle just below the posterior spiracles, or with some gall midges, they wedge their head or anal segments between ventral segments. The larva becomes a spring, it builds tension in its body by further contracting more muscles, concluding with one final wave that travels up through the body and is timed to coincide with the release of the attached parts, producing a massive recoil, propelling the little daredevil away. In 1992, David Maitland, from the University of Witwatersrand, determined that this way of getting about was not just a little bit better than wiggling, but at 0.5 m/s, a 200-fold improvement. This is fantastic way of avoiding being eaten or parasitized and, as Maitland points out, the flies are the only soft-bodied legless species (of arthropods) to use it.

It is the larvae of acephalic, i.e. 'headless' species, that we are the most familiar with in our day-to-day lives. These are the ones commonly referred to as maggots, and they are full of ingenious methods to enable movement – some are swimmers and have even developed flotation devices to aid them. In the foam nests (frothed saliva) of tree frogs, the larvae of seven out of the eight species of the genus Caiusa

(Calliphoridae) have been found living and feeding in the egg masses of the shrub frogs Rhacophoridae. In a paper published in 2018, Ananda Banerjee, a self-funded researcher from India, and co-authors discuss the behaviour of two species of fly found in eastern India. What they documented for the first time is quite remarkable. The larvae consume the hatched tadpoles until they are ready to pupate. They are found in frogs' nests, which usually hang over water and, once sated, they lower themselves on a thread into the water below and then swim to the edge. The maggot stays afloat using an air sac, while kicking in the water to propel itself along. I can't see it threatening the legacy of Michael Phelps any time soon, but this clever maggot is undeniably impressive. Sadly, Banerjee has told me in a personal communication that the habitat where this research was conducted is being destroyed at a rapid rate due to tremendous urbanization pressures. Just as we are discovering these creatures, we may be losing them and their many secrets.

The head capsule in maggots has been replaced over time by a so-called cephalopharyngeal skeleton, a series of hardened structures called sclerites that are usually divided into three main sections: the mandible, the hypopharyngeal sclerite and the pharyngeal sclerite. The mandibles are the mouth hooks, which are used for slicing food. In maggots, these hooks are distinctive and are sometimes the only way that we can identify a species. As well as feeding, the mandibles also play an important role in locomotion. Michael Roberts, an English entomologist published a paper in 1971 on the locomotion of Muscomorphan (neé Cyclorrhaphan) flies. He describes the movement of one of these, a larval bluebottle, *Callophora vomitoria*, as it crawls through a piece of meat, its standard food source. It first contracts and raises its anal part, and as this contraction passes through the body it moves, pulsing along like a caterpillar. This wave of rising segments passes along the body until it reaches the front end. The top of the front end, the cephalopharynx, is extended out of the body with the hooks raised up which are then driven into the meat to act as an anchor – think of it as resembling a modelling balloon that is squeezed from one end to the other. This movement through the meat is made possible by extra-oral digestion – yep, the maggot dissolves its surroundings and lubricates its path with a stream of saliva.

It is the maggot's ability to get into tiny and awkward spaces that has inspired researchers to think about how humans might move tiny things – for instance nanorobots – around in closed and remote spaces, such as in our body. This could inspire ways of moving drugs or repair materials to target certain specific areas. Bio-inspired mini-robots are not a new idea but navigating through biological systems is tricky. Not only are they tight spaces, how on earth are you going to keep precise control? Thankfully there are smarter minds than mine working at solving these problems. Chinese researcher Tong Shen and colleagues, engineers at the University of Colorado, Boulder, published a paper in 2017 on mag-bots. Yep, mag-bots, otherwise known as 'magnetically activated gel bots' based on the maggots of calliphorids and similar species. Now, robots using the telescoping motion of maggots described above have been made, but these devices have so far been relatively large. We're talking centimetres, and something that size moving around the body is going to hurt. But Shen and colleagues were working on millimetre-scale bots using a special gel remotely controlled by magnets. By oscillating the magnetic field, they could change the gel's temperature and so cause it to contract and expand. And they are now developing ways to use this to drive the mag-bot around in the same way a maggot moves. Tiny maggot-inspired robots may be exploring a body near you in the not too distant future!

Now, a flood of salivary juices may sound unpleasant, but we use these liquids in modern medicine to remove dead tissue, a technique called maggot debridement therapy. American physician William S. Baer is considered to be the father of modern maggot therapy, and was the first to observe that soldiers who had been injured in the field waiting for treatment with infected wounds, were less likely to suffer from gangrenes at a later date if their wounds were treated by applying maggots, which attacked and ate the dead tissue with some relish. He went on to conduct maggot experiments on patients who suffered from osteomyelitis, an infection of the bone that often resulted in dead surrounding tissue. A privately published autobiography, written in 1973 by Raymond Lenhard, who considered Baer a mentor (they both were at Johns Hopkins University School of Medicine), describes Baer's maggot work at Baltimore's Children's Hospital. His work was initially encouraging, but in the end these maggots resulted in the death of two patients. This prompted him to 'realise that he needed to create sterilized maggots'. He had up to that point been growing them on his windowsill in the hospital (it does not say what they were reared on).

You will be pleased to know that maggots are still in use, and that all are now sterilized. Maggots are also used as bio-weapons against pathogens. They are one of our last lines of defence in preventing ulcers and gangrene and against the effects of hospital superbugs such as antibiotic-resistant staphylococcus aureus (MRSA). Much is written about the wonders of bee vomit, aka honey, but it turns out maggot vomit is just as wonderful! As described earlier, maggots are able to mechanically deform and break down the outer membrane of the bacteria, but they can also secrete chemicals that denature that membrane. Mariena van der Plas is an Assistant Professor at the Leo Foundation Centre for Cutaneous Drug delivery, based in the Denmark. She has been working for a while now on novel biological approaches to dealing with skin infections and has focused some of her attention on secretions by *Lucilia sericata*, a Calliphoridae, whose maggots are firm favourites when it comes to debridement therapy. She and her colleagues found that the secretions oozed by maggots (not to be discussed at the dinner table) were actually invaluable to the recovery of patients with chronic wounds such as ulcers. Van der Plas and her colleagues found back in 2009 that the secretions of *L. sericata* were effective at breaking down the outer layer of both *Staphylococcus aureus* and *Pseudomonas aeruginosa*, arguably the most clinically important bacterial species when it comes to infections. If you sliced yourself open (a fairly common event for me as I'm a touch clumsy), the first phase of wound recovery is for the blood vessels to constrict, then for the platelets to bind together resulting in coagulation. The second phase is the inflammatory stage where the water, proteins, salts and so on leak out from the injured blood vessels and cause localized swelling. In most cases this is a good thing as it allows the repair cells to move around the wound and remove the harmful pathogens. However, when a patient has what is termed a chronic wound, this stage goes a bit haywire, and excess numbers of monocytes and macrophages, components that are normally beneficial, are produced in such numbers that they damage the tissue. Once more the magic maggots come to the rescue. Van der Plas and colleagues confirmed that the maggot secretions inhibited this excess production, while not impacting on any otherwise beneficial antimicrobial activities necessary for wound recovery.

This sounds marvellous, and so why are our doctors not regularly suggesting maggot therapy to treat our wounds? Well, apparently there is still a 'yuck' factor associated with them, they're not yet deemed a desirable option. This is remarkable when you consider that often, the only other outcome to avoid death is the amputation of the gangrenous limb. Fear not maggot chums, for there is a champion in the form of Dr Yamni Nigam, a lecturer at Swansea University in Wales, whose team is making it their mission to inform all who will listen about the fantastical properties of maggots, who they describe as 'cute, tiny little babies', while dispelling some of the myths. For instance, will the maggots turn rogue if let loose? No, they won't. With chronic wounds on the rise (we have no real figures about the global impact), this is a serious and often deadly subject. However, a paper published in 2017 by Krister Järbrink and colleagues at the Nanyang Technical University (and one researcher from a hospital in Sweden) reviewed reports from the USA and found that chronic wounds – often linked with the obesity epidemic – were affecting two per cent of the population with an annual cost to health services of $20 billion. Bring in the maggots, I say.

Many a maggot has incriminated a killer. When a body is discovered, one of the many things that the investigating team may do is remove any maggots or pupae from or around the cadaver. That's because the maggots, used in conjunction with other factors such as environmental and body temperatures, can give the police a time since when the death occurred – what is called the post-mortem interval. (Now, the perpetrators may have kept the body in a freezer and so artificially altered this timeline, and so the term minimum post-mortem interval ($_{min}$PMI) is the actual estimate preferred.) Flies, with their amazing abilities to sniff out a corpse (I discuss this in a later chapter) are generally the first to arrive at a crime scene, whereupon the mothers begin laying their eggs (or larva, depending on the species). We can see a less-horrific version of this each time we watch (Ok, I watch) decomposing wildlife in our gardens.

Blow flies (Calliphoridae) are some of these early colonizers, and we have been studying their larval development under different environmental and physical conditions for many years. Matthew Webb, a researcher at the University of Manchester and a keen maggot man but of the *Drosophila* variety, told me a story about a French student of his who was in the Gendarme. This student had wanted to establish the order of larval succession of bodies found in freshwater, and so, as many a forensic entomologist had done before him, he got a dead piglet, put it in a cage and submerged it. He thought about his design and realised that he was missing something – this was not similar to the bodies that they had been finding – and the missing component was clothes. To address this issue, he dressed the pig in one of his child's t-shirts. There are many complexities that need to be thought about when trying to design experiments to help us glean clues from a crime scene. Now though, we have more tools to help us, and we can ask even more questions of them thanks to advances in technology. In 2012, the first-ever case was broken using the contents of a maggot's gut. Maria De Lourdes Chavez Briones and collaborators, from the Autonomous University of Nuevo León and The State Institute of Research in Mexico, recalled how they used maggots to identify a very badly burned corpse. After a girl had gone missing, the police had found a corpse that they could only very tentatively identify as the missing female – they had found a school graduation ring on the body. But the identity could not be confirmed because the body had been so disfigured and was already badly decomposing. Chavez and her team hypothesized that maggots found on the body might be able to determine the identity of the body, as their stomachs would probably contain small but identifiable strains of DNA from their host. And it turned out, they did. DNA later found in the maggot's stomach was confirmed to be a familial match with the girl's father.

And what goes into the stomach must come out. Even the waste products of a larva can be useful, and once more it is the chironomids who are helping out. For instance, they can help purify our water. Roger

S. Wotton, Emeritus Professor of Biology at University College London, and Kimio Hirabayash, a Professor in Shinshu University in Japan, published a paper in 1999 looking at the amount and value of maggot faeces in water purification treatment plants, where sand is used to remove impurities. The midges eat and live in what is, amusingly for me, called the schmutzdecke layer – the German for dirt cover or dirty skin. It is this layer where most of the water is purified, as it is rich in bacteria, fungi, and, along with other micro-organisms, these midge larvae. The midges ingest a lot of food and so produce a lot of faeces. The authors state that 23% of the surface area of the sand in their investigation was covered in the stuff! This hugely increases the surface area of the sand and adds to its structural complexity, enabling more organic material to bind to it thus trapping it in the sediment, further adding to the purification process as it no longer leaches out. Oh, the power of poo.

Once the larval stage ends, one of the most miraculous events in nature happens. Flies are now at the period in their life where they undergo a change that all holometabolous insects go through – the name holometabolous deriving from the process of complete and often extreme metamorphosis. Many insects, including the flies, do not just grow and develop in stages but instead a whole new development stage, the pupal stage, has been added. It is within this stage that the organs and tissues are reorganised, remodelled and sometimes, rebuilt, a drama about which we still have many questions.

Charles Darwin writes in his book *The Voyage of the Beagle* about a German collector called Renous whom he met in Chile back in the 1830s. He recounts a story Renous told him, several years prior, of some caterpillars that Renous had left under supervision, at a house in San Fernando, Chile. During his absence, they metamorphosed into butterflies, which the local community, including the Governor and local clerics concluded must be heresy, and so arrested him. I like this extreme response to ignorance, but I concede it is not helpful in furthering scientific knowledge. The Egyptians were the first to link worms and grubs to the adult insects, but to really see what was happening in detail, we needed a closer way of looking at things. We had to wait for the 17th century and the invention of the microscope for that. At this time it was determined that larval, pupal and adult insects were all linked but it was Dutch microbiologist Jan Swammerdam, in 1669, who was the first to show that they were not separate creatures as it had previously been argued.

We know now that the adult structures arise from little groups of cells called imaginal discs, which first start to develop way back in the egg stage. Swammerdam could easily see this in Lepidoptera, the butterflies and moths, but struggled with the flies, as they did not take the shape of the adult structures, unlike those in the Lepidoptera. Swammerdam had cut open silkworms to show the beginnings of the wing developing below the surface but could not find similar structures in flies, and that's because these mostly remain dormant until the pupal stage. He further recognised that different insects metamorphosed in different ways. Today we recognise three groups – ones in which none or very little change occurs, with the adults being wingless (ametabolous); ones who look like mini-adults on hatching, with the exception of wings and genitals which develop in stages through life (hemimetabolous); and the ones that undergo a complete change (holometabolous).

Most insects undergo complete metamorphosis, and there are pros and cons to this strategy. Jens Rolff and co-authors propose in their 2019 paper that this enables insects to exploit short-lived resources. If you have had an exceptionally good week feeding and have enough resources to change, why wait? Being able to exploit different resources at different stages in their lives, as well as developing structures to further particular roles, such as mating and dispersal, has enabled the massive explosion in species numbers. There are costs. Hiding yourself away makes you defenceless against predators, parasites and pathogens. My favourite example of protection is within the Family Phoridae, specifically the genus *Pseudacteon* – the ant decapitators. The larvae of this genus are parasitoids in ants, resulting in the head either falling off or being cut off. A totally gruesome result for the ants but an exceptionally clever one for the fly, as it now has a really tough, protective environment in which to pupate.

How this process of metamorphosis evolved has not been properly resolved, although work by American biologists James Truman and Lynn Riddiford has greatly improved our understanding by proposing a

An adult *Pseudacteon* sp. (Phoridae) bursting out from a decapitated ant's head.

comprehensive theory. Insects that undergo incomplete metamorphosis go through a very short stage – after the egg but before the nymph (larval) stage, called the pro-nymphal stage, that may be free living or still within the egg. And it is with this stage that Riddiford and Truman propose that things started to happen, as they suggest, by a chance mutation over 300 million years ago (most of life is a chance mutation so this is not as unbelievable as you may first think). Baby birds – the chicks – absorb all the nutrients whilst in the egg and then hatch as an adult. But what if some of the resources remained unabsorbed, what if some of the insect pro-nymphs were about to hatch but there were food supplies still left? Not to be wasteful, these pro-nymphs although ready for life on the outside, decided instead to stay inside the egg. With time, it is thought that this evolved with them leaving the egg but continuing to feed in a similar fashion, and as such still staying in this proto-nymphal stage. Eventually, these little creatures evolved into the larvae that we know today, often feeding on a completely different food source to the adult. The pupal stage is proposed to be a condensed stage at the end of this that enabled a dramatic change into adulthood.

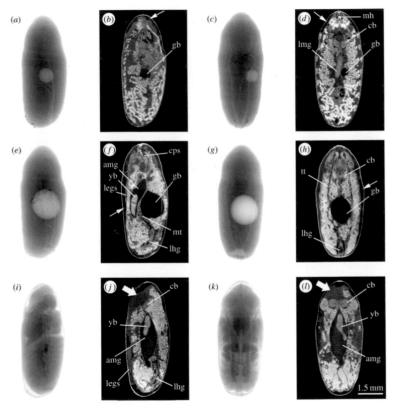

Lateral x-rays (left) and micro-CT (right) images of the pupal stage. The first row is 6 hours since the pupa formed, the second is 24 hours and the third is 30 hours. This is just the first 0.5% of the pupal stage. The fat arrows show the beginning of the adult head. (amg, adult midgut; cb, central brain; cps, cephalopharyngeal skeleton; gb, gas bubble; lhg, apoptotic larval hindgut; lmg, larval midgut; mh, mouthhooks of cephalopharyngeal skeleton; mt, malphigian tubules; tt, tracheal trunks; yb, yellow body).

This incredible process is all down to two regulatory chemicals, ecdysone and a group of juvenile hormones, the former controlling moulting while the latter ensures growth while inhibiting metamorphosis. We have learned to not only synthesize products to affect these but also now are starting to fully understand the genes that are producing the hormones, and so disrupt the lifecycles of insects that we consider pests – an example is with mosquito suppression by methoprene that acts as a growth regulator.

What physically happens during the pupation stage is also being revealed finally thanks to some exciting new imaging techniques, such as microcomputed tomography (micro-CT) scanning. Daniel Martin-Vega, a post-doc student of Martin Hall, at the Natural History Museum, London, has been exploring ways of using imaging systems to study larval and pupal development for many species of fly, especially the forensically important species that aid many a police inquiry. Researchers have been working on the larval stage for a while and have built up comprehensive accounts of larval development in environmentally different conditions. But they have struggled with the pupal stage, as its development is hidden. This was a mystery phase during which we had no clue as to what was happening when. Hall has been leading a team of researchers for a while, including Martin-Vega, that is using micro-CT scanning to image what is happening inside the pupal case during this phase of development. And it's amazing, with the fly undergoing many bursts of development rather than one continuous one. Martin-Vega carried on playing with this technology and was able to show age-diagnostic characters at 10%-time intervals during the pupal development, termed the intra-puparial period. The police now have a very good method for estimating the post-mortem interval.

We have only really just scratched the surface to understanding the larval and pupal stage of flies. Most are ignored as they are quite difficult to identify, or rather more importantly, hard to find! So many species are yet to have their larvae described. Can you imagine what gems of information are waiting for us?

CHAPTER 2

Heads up

Since everything is in our heads, we had better not lose them.

Coco Chanel

TO HUNT, MATE AND EAT we must first see, smell or hear. We have to process and understand information from our surroundings, and we have to act on this information once we have understood it. It may be good information, a delicious meal of a freshly produced dung pat, or it may be bad, coming face to face with a wasp intent on using you as a living larder for its young – not a fate you would choose. Whatever the reason flies, as with all animals, need to use their head and make up their minds rapidly, and indeed most of a fly's sense organs associated with sight and smell are located on the head. There have been some extraordinary modifications within the order when it comes to sense structures (both internally and externally), enabling them to live and breathe another day. I am leaving out the mouthparts and antennae from this chapter as too many modifications have happened with these structures and they deserve their own chapters.

When it comes to describing the different species of fly we use many characteristics associated with the head. Once an individual fly has undergone metamorphosis inside the pupa, it is ready to leave and embark on the final and most fun stage of their life cycle. But how to

Mosquito bobbing. The mosquito larvae dangle down the water column taking in their surroundings.

The puparium, the emerging adult, and the head showing the ptilinal ridge surrounding the antennae of the common house fly, *Musca domestica*.

get out of the pupa home? Many of the nematocerous species have very thin pupal cases that are easy to escape from but in the Brachycerans the pupal case is thickened and very tough. They have developed some interesting techniques. One interesting group in this regard is the higher Muscomorpha – the Cyclorrhaphan flies, a name derived from 'rhaphe' meaning seam, along which the puparium splits allowing the adults to emerge. The two main groups in the Cyclorrhapha are the Aschiza and the Schizophora, and these have two different ways of doing this. The hover flies are in the group known as Aschiza, a name that means 'without split/cleavage'. This name was coined in 1882 by Austrian Dipterist Eduard Becher who was the first to classify them into what we have since determined as a group, defined by what they don't have rather than what they do.

The Aschiza just push through the weakened seam, but the Schizophora have an added structure to facilitate this action – they have a protrusible bladder called a ptilinum which acts like an inflated air bag. They use this sac to pop the head off the end of the puparium, enabling them to flee their teenage home. The sac then deflates back into the head leaving a distinct ridge that frames the mouth parts – a key diagnostic feature for adult Schizophora. The Schizophora are further divided into the acalyptrates and calyptrates (see Chapter 6, p. 161).

The shape of these all-sensing heads varies in flies, with two key forms: prognathous, where the head follows a horizontal plane and the mouthparts are at the front, and hypognathous, where the head is more vertically aligned with the mouthparts angling downwards. The former is much less common and is found in the crane flies. This common name applies to the four groups of flies of the superfamily Tipuloidea. We can't use the term family for these as they are taxonomically problematic. Traditionally, Tipulidae comprised four subfamilies, which were then promoted to families: Cylindrotomidae (long-bodied crane flies), Limoniidae (short-palped crane flies), Pediciidae (hairy-eyed crane flies – I kid you not) and Tipulidae (true or long-palped crane flies). A 2010 publication by Matthew Petersen and co-authors, at Iowa State University, suggests that there are just two families – the Pediciidae and Tipulidae (containing as well Limoniids and Cylindrotomids). But whether this is correct, i.e. a true representation of their phylogenetic relationship is still being debated, as is the second ongoing debate of how these flies relate to other flies. This latter point is of interest here as prognathous heads are considered ancestral and there is a lot of work studying the position of mouthparts in hemimetabolous insects. As far as I can determine, no such research has been undertaken on the flies. Traditionally the tipuloids were thought to have been one of the early lineages but now we think this may be inaccurate, and that they evolved much later. This debate has been going on for over 100 years and may go on for a few years more.

Whatever the final taxonomic outcome is, the adults of the superfamily Tipuloidea, especially those in family Tipulidae, have very distinctive heads. Bulbous eyes sitting on top of a pronounced snout, their heads superficially resembling that of Pluto, Mickey Mouse's pet hound. Not all adult crane flies are concerned about feeding – some don't feed at all – but

The forward-pointing head, with its dangling maxillary palps, of a *Tipula* sp. (Tipulidae).

focus rather on reproducing and dispersing to new habitats. They often have long, dangling maxillary palps, which assist with feeding, but are also thought to be useful for sensing chemicals emitted by the opposite sex. Those adults that are short lived need all the tools they can use to see, smell, and track an appropriate suitor, to make sure their genes are passed on. More commonly flies have a hypognathous head, with mouthparts facing downwards, with their exact features varying enormously across and within different fly families.

Many people think adult flies lack brains. How could animals so small think and learn? Surely they have only instinct, with little real 'thought' governing their behaviour (sounds like my mother telling me off!). Watching them fly straight into windows again and again only reinforces this misconception. But they do have a brain – and a remarkable and very useful brain at that. Consider once more *Drosophila melanogaster,* quite the darling of the research world. This diminutive

The downward pointing head of a blow fly, *Calliphora vicina.*

3 mm-long creature is at the forefront of research into the development and function of the nervous system. This species has been helping us understand genetics ever since Thomas Hunt Morgan (1866–1945), set up the 'Fly Room' at Columbia University in 1908. He was fascinated by genetics and was the first to receive a Nobel Prize associated with *D. melanogaster* in 1933 for his work on the role that chromosomes play in heredity. After a century of studying this simple brain, there are still many unanswered questions, but new understanding comes daily.

Drosophila brains are described as plastic – they are able to cope with changes and modify themselves after each experience or stimulus. Childhood experiences count for a lot, even in flies. What I find amazing is that this can happen at any time in the life cycle – an experience as a larva will impact on how the adult later behaves; they learn and adapt. An example of a fly's ability to use its brain involved an experiment in 1989 in which researchers Karl Kral and Ian Meinertzhagen, at Dalhousie University, removed the front legs of flies – the limbs that are used to clean the precious sensory equipment on the head – to see what would happen.

Flies are not the dirty creatures they are commonly perceived to be – in fact they are fastidious cleaners. They need to keep themselves clean, free from dirt, debris and pathogens. The cleaning of each part of the body is organized hierarchically – with the most important parts needing to be cleaned first and more regularly – eyes, antennae and head are cleaned first, and then they move down the body cleaning each part with fastidious attention. After a couple of days, the flies without legs had twigged that they were lacking front legs, and so they started cleaning themselves with their middle set of legs. They had thought the problem through and solved the dilemma. How flies are processing all of this information is well out of the realm of this book (and my own brain), but there is a lot of research going on right now that is trying to figure it out.

Flies, like all creatures, rely on their senses to guide them towards food and mates, and away from predation. In 2018, a breakthrough with imaging helped this process, with the beginning of mapping of the *Drosophila* brain at the nanoscale level, that is, at a scale of 10 millionths

Flies are not filthy, quite the opposite, they are always cleaning themselves.

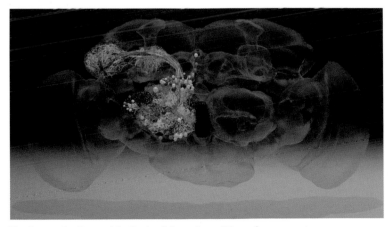

The brain of a *Drosophila*. Each of the coloured threads represents neurons, 100,000 in total which can be traced from start to finish.

of a centimetre. American Davi Bock and his team at Howard Hughes Medical Institute's Janelia Research Campus, took 21 million images of a *Drosophila* brain, stacked them together and produced an incredibly high-resolution image.

Bock and his team used a process that rapidly takes many images and mashes them together. You may think that taking 21 million images and processing them would take a few millennia, but they achieved this in less than seven minutes, thanks to an innovative imaging system and stacking software. From this they were able to complete a synapse-level reconstruction of the *corpora pedunculata*, commonly referred to as 'mushroom bodies' due to their shape. These, along with the lateral horn (lateral protocerebrum), are the areas of the brain that receive the stimuli – the olfactory information from smells, via the antennae. The mushroom bodies appear to be pivotal in establishing links between the stimuli, in simple terms, how it learns. Bock's maps can help us manipulate cells and genes and thereby learn more about brain function. This ability to learn is not restricted to *Drosophila* but is taking place in the brains of every fly on the planet as you read (and so in some humans as well).

Understanding the brain also has some fun transferable applications. One such example is being researched by Professor Holger Krapp, at Imperial College, London. The human brain has 86 billion neurons, specialized cells for transmitting impulses. But just because it is bigger than a fly's brain, which on average has 100,000 neurons, does not always mean it is better. Even though the fly's brain is small, its reactions are incredibly fast – try swatting a fly for example, it more often than not evades your actions. (Don't actually do this though as it's cruel.) The number of neurons in the insect brain may be fewer than an human brain, but the processing speed is phenomenal. Krapp and his research group work on 'comparatively simple insect model systems to discover fundamental principles of task-specific sensory integration, multi-sensor fusion, and sensorimotor transformation'. He is using a fly's superfast responses as inspiration for developing faster supercomputers to run, for example, sophisticated aircraft. Specifically, his team are looking at the imaging systems of flies and how they communicate the responses needed to deal with stimuli to their wings.

Flies are the masters of flying. And one of the ways that they are such great aviators and don't crash into everything is by using optic flow (OF) clues, which enable them to process thousands of images at speed as they move around. Having just one type of eye, like humans, is not the fly's way. They are not just the masters of the air, but also of the land, often hidden away inside caves, the soil and bodies, and need to be able to effectively get around in these habitats. Flies, during their life time, have different 'eyes' to facilitate this. For the adults there are the compound eyes made up of ommatidia (the photo-units), the ocelli (simple eyes) and the Hofbauer-Buchner (H-B) eyelets. The first two eyes are familiar to many but the H-B eyelets were only formally identified in 1989 and their origins determined in 2002, plus they are very small. To understand where H-B eyelets came from, we need to first look at the larval eyes.

I didn't mention how a larva visualizes its surroundings in the previous chapter as some of the features are linked to the adults – there is some sharing between the larval and the adult stage, which is unusual for external features such as the larval eyes as they are lost during metamorphosis. The larval eyes, the stemmata, are derived from compound eyes (all immature stages of hemimetabolous insects have compound eyes which they grow during their life span; the holometabolous have gone the opposite way and have reduced eyes in the larval stage), and these vary in their complexity. American entomologist Cole Gilbert, in 1994, wrote that the Diptera have 'perhaps the widest range of stemmatal morphology among the holometabolous orders'. Why am I not surprised?

The stemmata in nematocerous species and lower Brachycerans, comprise of photosensory cells with lenses, referred to as larval ommatidia, and the number of cells varies, with numbers ranging from 40 to 45 down to 7 or 8. The larval eyes, if external, lack the corneal lenses so important for the adults, and instead have simple crystalline cones under a transparent region of the cuticle. The larvae can detect changes in light and dark, and most of the time that is sufficient

Some of the aquatic species have opted for a little more bang for their buck. Jan Swammerdam in his *Bybel der Natuure* that was published in

1737, writes about the eyes of the larvae of a culicine mosquito. He found these to be working eyes unlike those of other larvae. We now think that what he was describing were the highly modified eyes that we know are found in both Culicidae and Chaoboridae (phantom midges) – the latter family has some seriously predatory larvae (see pp. 36–37).

Unlike the rest of the larval ommatidia, these unusual 'eyes' lack the corneal lens and have a eucone crystalline cone, both features seen in adult compound eyes. These flies have both the normal larval eyes, the stemmata, but they also have these 'compound eyes' that have good visual functioning – both these larval eyes and adult eyes contain a rhabdom (the photoreceptive area) comprised of retinula (photoreceptive) cells – essential for hunting and avoiding being hunted. Nothing is going to get past these flies.

Some larval eyes don't have any lenses, their stemmata have been reduced to a collection of photosensory cells below the cuticle, a process that has been greatly extended in many of the muscomorphans. The larvae of these flies are found in the soils, in trees, in leaves, in caves, in fruit and in animals' nostrils and, as such, most don't have the need to see as clearly as their adult counterparts; their obliging mothers have laid them in habitats that already surround them with their food. In these larvae, the photosensitive cells have become completely internalized back into the head and have clustered to form a pigment cup of 12 photoreceptors called the Bolwig's organ.

Named after Niels Bolwig, a Danish entomologist/biologist, he was as diverse as he was

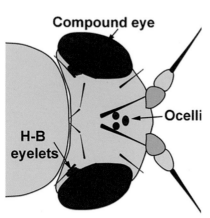

Not content with two, flies have up to seven eyes, which consist of two compound eyes, three ocelli and the modified Bolwig's organs (Hofbauer-Buchner or H-B eyelets).

talented working on many different subjects, both topics and animals – the title of his autobiography, published in 1988, was *From Mosquitoes to Elephants*, pays credit to this. Bolwig, in 1946, was the first to write about this group of photoreceptors pocketed in the head of house fly maggots. Bolwig's life is something to aspire to but sadly mostly ignored. Not only did he conspire against the Nazis in his homeland, he was later forced to leave South Africa because of his stance against apartheid in 1959, he then was President of Nigeria's Wildlife Preservation Committee and wrote the Nigerian Wildlife Protection Laws, he went on to teach all over the world up to 1979 when he eventually retired to Somerset (a lovely county in the UK and I can understand that decision). All that and the man only got a Wikipedia entry in 2020. He is best remembered for his work on maggots and their inner eyes.

In *Drosophila*, the Bolwig's organ is composed of just 12 photoreceptors but in 2010 scientists found out something rather interesting – if you remove the Bolwig organ, the larvae will still respond to light. Yang Xiang and colleagues, from the Howard Hughes Medical Institute at the University of California, were playing around with light receptors in a *Drosophila* larva and found, much to their surprise, that they recorded neurons that responded to ultraviolet, violet and blue light, and are important in avoidance mechanisms, particularly at high intensities i.e. when the larva nears the surface of a habitat that may be dangerous for both predation and desiccation. This was not the novel part of the study as we knew this about larval eyes – what was new was that this was not just found to be the case with the eyes but across every single body segment of the larva. The idea of different parts of your body being able to sense changes is an incredibly useful one when you want to avoid being eaten.

The stemmata of all the dipteran larvae go through a radical change during the final larval and pupal stage. *Psychoda cinerea* is a cosmopolitan species, within the family Psychodidae, the moth flies; the adult is incredibly hirsute and holds its wings parallel to its body, so as to resemble a fluffy-caped crusader. It was not the adults that were of particular importance, but the larvae. Dr Peter Seifert and colleagues, at the University of Cologne in Germany, in 1987, were the first to see and document the larval eyes being contracted back into the head (these were originally external), and then new, larger, more developed eyes, the adult compound eyes, grow in their place, eventually shoving the larval eyes back into the head.

This not only happens with the external eyes but also with Bolwig's organ. In 1989 Alois Hofbauer and Erich Buchner, both at the University of Würzburg, published a paper describing the H-B eyelets in adult *Drosophila*, which they found to be important in regulating the fly's circadian clock. It seems quite remarkable that we have been extensively studying this genus for so long and missed these little structures! They speculated then that these may be formed from larval lateral ocelli (they are not in fact ocelli). But it took 23 years for researchers to prove this. In 2002 Charlotte Helfrich-Förster and co-authors reported that the eyelets had the same biochemical response as the cells in Bolwig's organ, showing a link between the two structures through metamorphosis. The larval 'eyes' take their long-unnoticed place squashed next to the highly sophisticated adult compound eye.

The remarkable visual capabilities of the compound eyes that the adults have, has slightly overshadowed the ocelli, but both of these are important. Ocelli are referred to as simple eyes, as they are made up of a single lens, whereas compound eyes have multiple lenses. Ocellus literally means 'little eye', and when present, they are located on the top of the head, a region called the vertex, which surrounds the ocellar triangle, with one ocellus at the front and two behind. These little eyes are very useful for us in classification systems, but there is a lot of variation across flies – some families uniformly lacking them, in other families only a few species or groups of species lack them. Still others only have the front ocellus or only the lateral ocelli, and sometimes not even that but just an ocellar tubercle (the raised-up section of the triangle).

Ocelli are basically poorly-formed cameras. Blurred blobs are the best sort of images flies can hope to attain from them. The ocelli are not used for sharp vision, instead, they are light sensors. The two ocelli at the back and one ocellus in front provide a picture created by overlapping light levels. The ocelli act as a spirit level – if the fly is knocked sideways off its flight axis then it can correct its position; the images provided by the ocelli help it straighten itself. The compound eyes can also do this, but not as quickly, as they have so many more images to process.

The most complex eyes, the compound eyes, have been around for millions of years and are a feature of all arthropods not just flies – trilobites that date back 540 million years have them. Each compound eye is composed of a number of individual units called ommatidia. These units contain clusters of photoreceptor cells – the retinula cells – that are enveloped by pigment and support cells. These retinula cells vary in number in different arthropods. For example, the butterflies and moths have nine (R1 to R9, but R8 sits on top of R9 and so just counts as one), but orthoptera (the crickets and their relatives) have 8 (R7 and

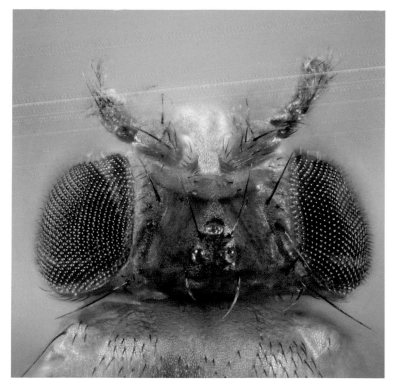

The location of the ocelli and ocellar triangle (vertex) on a fly's head.

R8 again have this two tier system). These cells are arranged so that their light-gathering surfaces are fused at the centre of the ommatidia to form a light guiding tube called a rhabdom. This surface is called the rhabdomere and it is covered with microvilli (loads of little finger-shaped protrusions).

The rhabdom is where the light-sensitive pigments, including rhodopsin, the pigment that enables insects to fly at night, are concentrated. Around 100 million years ago the flies decided that this was no longer good enough for them. They didn't want to keep adding more and more ommatidia to sharpen the picture, the dragonflies can keep that method. Flies instead, separated the rhabdom, so that the seven retinula cells (R7 and R8 are again fused in flies) no longer have connecting rhabdomeres.

Each ommatidium in an adult fly consists of a lens on the surface of the cornea, below which is a gelatinous area called a pseudocone which is the focusing apparatus. This is surrounded by a of row of semper cells below and primary light pigments in the side walls which prevent light entering from the neighbouring ommatidia.

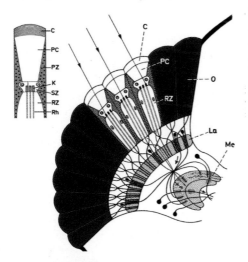

The retinula cells R1 to R6 are used for motion vision, and contain rhodopsin, that is able to detect a broad range of wavelengths. The axons that connect these cells to the lamina, the part of the brain that is specialized for motor vision, contains information not just from this ommatidium but

A single ommatidium and its arrangement to form the compound eye of *Musca* sp. (Muscidae).
(C, corneal lens; PC, pseudocone; RZ, retinula cells; PZ, pigment cells; K, rhabdomere cap; SZ, Semper cells; Rh, rhabdomere; La, lamina; Me – medulla.)

they are combined with six further rhabdomeres from the surrounding ommatidia that share the same visual axis – producing a finely detailed image. The tiered R7 and R8 have different rhodopsin's that are much more restrictive to certain wavelengths and in making comparisons in this restricted spectrum they are able to determine small colour changes. The axons for these cells head off to the medulla, another part of the fly's brain. Due to having a full complement of neurons connecting cells from different ommatidia, cells from the same ommatidia and a whole lot of moderating neurons that process and filter much of this information, a whole lot of rewiring in the eyes neural network has occurred.

A fly's eyes are the natural equivalent of a photographic stacking system and a network of nerves pass the information on to the processing centres, called the neuro-network. The processing of these images in the brain has become super enhanced too, as researchers such as Krapp are trying to discover. The ability of flies to recognise and respond rapidly to what they see enables them to successfully enter hostile environments, such as your garden, with all its hidden threats, and come out alive. They possess skills that far surpass those of vertebrates (the birds and the bats etc), skills we are trying to mimic for robots, both on the ground and in the air, and intriguingly self-driving cars – the ability of a fly being able to dodge objects coming from all directions at all speeds would be handy in this latter capacity. So, although dragonflies are thought to have the most impressively complex eyes – they have up to 30,000 ommatidia – flies have a greater sight sensitivity without having had to increase the size of their heads or reduce the sharpness of the image.

The number of ommatidia depends on the species – some have thousands whilst others don't have any at all, as in many a cave-dwelling species. In 2016, a paper published by Trond Andersen and collaborators found a truly remarkable pecies of chironomid. At a depth of 980 m (3,215 ft) in a cave in Croatia, they found seven females of a type of non-biting midge that they subsequently named *Troglocladius hajdi*. Both the genus and the species were new to science. The genus name translates as 'cave midge', while the species name pays homage to the winged underground fairies from Slavic mythology. No males were found. This led to a theory that this species may use asexual reproduction, or parthenogenetic reproduction, whereby the females clone themselves. The eyes (not all had eyes) were tiny, hairy, kidney-shaped structures with just a few ommatidia due to the midges living in total darkness.

Troglocladius hajdi really are incredible creatures with their tiny eyes, but it is more common for flies to have loads of ommatidia – *Drosophila* have 800, *Sarcophaga dux*, a flesh fly, has on average 6,032, and *Syritta pipiens*, a hover fly, has 6,400. However, we've only counted the facets on a small number of species so who knows what the largest figure is? Although not individually counted, many have now been mapped and in the process, we have come to realise that not all areas on the eye are the same. Some areas are termed 'acute zones' as, due to very small angles between neighbouring ommatidia, an incredibly high-resolution image can be developed. Other areas are called 'bright zones', where there are larger ommatidia for greater sensitivity. There is also a 'love spot' – honest, that is the proper term for it – in which males, of *Musca domestica* for example, have more, much larger, ommatidia. It is a section of the eye that specializes in hunting fast-flying objects – in this case, the ladies. The eyes of male and female Bibioniidae are very different from each other. In fact, the heads of the two sexes are so different that they are often confused as two different species. For the males have beautiful bulbous, ommatidia rich, bulging eyes. And these have been divided into two sections so that he has now two dorsal and two ventral eyes. The dorsal eyes have larger facets and longer rhabdomeres to give him greater long-distance vision.

Male *Syritta pipiens*, the thick-legged hover fly, found across the temperate regions of the world, can see females before the females see them! Thomas Collett and Michael Land, both at the University of Sussex, published a paper in 1975 on the flight characteristics of this species. Hoverflies are some of our favourite flying flies, but we miss so much about their crazy behaviour because many of their rapid movements are invisible to the human eye. Collett and Land were able to study these by first filming them, and then studying the behaviour frame by frame. At first many *Syritta pipiens* individuals paid no attention to each other and just cruised on by. But if the attention of the male was piqued, he would start tracking the subject; firstly, by turning his body

The hairy and bulbous eyes of a male *Bibio marci* showing the many ommatidia used by the males to search out females in a swarm.

towards it, and then jerking rapidly to ensure that he had a correct fix on it, building up a mental map of its exact location. The males have a region of enlarged eye facets at the front of their eye, two to three times the resolution of the rest of their ommatidia, and this enables them to see and track their target very effectively. He keeps a constant 5–15 cm (2–6 in) distance from his target and will adjust if the target changes course. Amusingly if two males are on the same flight path heading to each other straight on, they will 'wobble' themselves apart, e.g. they make little left-right movements off their original path where the distance of the left flight is slightly longer than the right, and so gradually moving away from the collision line. Once the object of the male's attention settles, he either arcs around it or, at an astonishing acceleration of 500 cm s^{-2}, he darts towards it, with his genitalia at the ready, to, ahem, strike. The authors note that the male engages in this sexual activity with 'either sex indicating that successful copulation involves more trial and error than recognition'.

It's not just about sex; these alterations are useful for hunting. Take the robber flies (or assassin flies) – the family Asilidae – for example. These are some of the fiercest and most successful predators on the planet. Mostly ignored in TV series about predators (it sends me into a social media frenzy), these fearsome beasts deserve to be seen as the true barons of the sky. Back in 1983, Anthony Joern and Nathan Rudd, both then at the University of Nebraska, published their findings on robber flies hunting grasshoppers. They were studying the impact that *Proctacanthus milbertii*, one of the most widely distributed robber flies in North America, had on the resident, and physically much bigger, grasshoppers of the Arapaho prairie. Joern and Rudd estimated that the population of grasshoppers was about 64,000 per hectare and the predation level ranged from 0.5 to two grasshoppers a day per robber fly. That gave a percentage kill rate of 0.5% to 2% of the adult grasshopper population in a day – a significant amount. These are solo hunters, feasting on grasshoppers that are more than double their size – they are feeding on the dissolved insides of the grasshoppers and not the bulky exoskeleton (in mammals, it is very unusual for solitary animals to feast on creatures larger than themselves, most are of similar or smaller size,

The eye facets of the robber fly, in the genus *Holcocephala*, showing the larger central ommatidia and the more numerous smaller ommatidia surrounding them.

and even if this is the case, they are never as prolific as these flies). Their incredible vision in this case is combined with another fantastic feature, their venom (which I discuss in Chapter 4).

It's not just the large robber flies that are formidable hunters – the largest species within this family is a species from Madagascar, *Microstylum magnum*, commonly called the giant robber fly, which reaches an impressive 6 cm (2¼ in) long, with an 8.4 cm (3¼ in) wingspan. At the other end of the spectrum is a robber fly the size of a grain of rice, which shows that size isn't everything when it comes to being a kick-ass predator. An international team of researchers led by Dr Trevor Wardill, at the University of Cambridge, in the UK, in 2017 published their research on *Holcocephala fusca*, a species from North America, a 6 mm (¼in) – long predator that they nicknamed Top Gun because of its incredible ability to catch prey. These flies work out the trajectory of their insect meal and then intercept them on their line of travel, in much the same way we catch a ball while running. These stunning little flies have greatly enlarged ommatidia at the front of their head and very extended rhabdomeres (same as the love spots) which opens up the area that they can focus on, and so can accurately see objects at a considerable distance – my, what big eyes you have Mr Robber Fly.

At exactly 29 cm (11.42 in) away from its target – a very specific number and no one yet knows why this is – it changes both its direction and its speed at angles and speeds that have not been described for any other flying animal. Initially, they fly full speed, but at 29 cm they slow down and approach their victim from behind. Wardill compares this technique with the way relay runners pass batons to each other – when they are parallel to each other it is considerably easier to pass the baton than when they are running at each other. These tiny flies are supremely efficient aerial muggers, and their abilities to do so on a small scale are being studied to improve the efficiency of aerial drones. The speed at which a drone can process images, and the time that it can remain airborne, is dependent on the battery. If you can increase either processing time or time needed in the air, while also reducing the size of the battery, you could be able to enter incredibly small environments or cover long distances.

The amazing patterns on the eyes of horse flies. This is a *Chrysops*, which is the Latin for gold eyes.

The eyes of flies come in many forms and fashions. The red of the eyes of *Drosophila melanogaster* is caused by pigments in the ommatidia that absorb light of wavelengths <600 nm. Red and green are the more common colours in flies, but there are many other multicoloured varieties. Some families have reflective colour patterns, which make flies eyes some of the most beautiful of all animals. There are stripy eyes, spotty eyes, and seriously, where else do you find eyes that have triangles of colour on them? Some have distinct patterns on them, as exemplified by the family Tabanidae, the horse flies.

These patterns are not caused by colour pigments but by corneal colour filters that partially reflect light. Many of the patterned eyes have nipples on them – corneal nipples – and these can be seen in many families including the horse flies, mosquitoes and crane flies. These tiny cone-shaped protrusions, 30 nm in height in *Drosophila*, alter the refractive index, or more simply put, how fast the light penetrates the eye. White light is composed of a spectrum of colours, and by refracting it, we separate the colours in a process called dispersion – in flies, especially the horse flies, this results in some very pretty colours and patterns. When the fly dies, these patterns die as well, not because they fade (there is no pigment to fade) but rather because the eye slightly collapses, causing it to deform, and so the structures that originally caused the colours are altered.

In horse flies, these nipples are much larger at nearly 70 nm, and the single nipples have merged into what appears to be a maze. This arrangement of nipples, called nanopatterns, in the eyes of flies are thought to be a fantastic replicate of Turing patterns. Turing patterns were first described in 1952 by the English mathematician Alan Turing (who also helped crack the Enigma code in World War Two) and are a type of reaction-diffusion model. These models describe the patterns formed as different chemical substances diffuse across a surface – or through an embryo – and react with each other as they come into contact. The key feature of Turing's model is that tiny, random disturbances can cause an initially symmetrical system (like a spherical embryo) to form new, more complex equilibria – stripes that are seen on the eyes of tipulids (crane flies) and the mazes of tabanids. In fly ommatidia the patterns are caused by organic chemicals on the surface of the lens reacting with each other. When Turing published his paper *The Chemical Basis of Morphogenesis* nearly 70 years ago it was purely theoretical, but since then we have realised that these patterns are everywhere – from the stripes of the zebra fish to the spacing of mice hair follicles – but only across the eyes of arthropods do you get all possible variations (from dimples to nipples to mazes and everything in-between). So why do flies such as tabanids have these patterns? What benefit do they bring to the adult fly? By altering the spectral composition of the light, and causing colourful reflections, we think this brings an advantage in mate finding and subsequent courtship, and also in finding food as the nipples increase contrast sensitivity when comparing information that is gleaned from different ommatidia with different filters.

Females in particular have very colourful eye patterns. Amazingly, these patterns can help flies fade out the unimportant colours such as green from the environment to make the objects that they want to identify – the large dark bodies of their blood supply – for example, become more pronounced. Nature has found a way to counteract their blood lust though. A marvellously bonkers paper was published by Tim Caro and co-authors in 2019, in which they disguised cows as zebras. This sounds like fun in its own right, but there was a scientific reason behind this – they wanted to see if the zebra patterns affected the feeding habits of flies. They found that the cows wearing coats of many colours or stripy coats were attacked less than cows dressed in either black or white coats. All cows had uncovered heads, that acted as a control area for that cow, and irrespective of coat that cow wore, the rate of attack on the head was the same for all cows. It was the coat and not the cow that was affecting the biting rate, with the patterns obstructing the fly's ability to work out where the cow was.

Most adult flies are able to move their heads and therefore their eyes around with astonishing mobility – Krapp and his lab filmed a hover fly rotating its head 270 degrees, which is the same as the rather more famous head-turners, the owls. Humans can just about manage 45 degrees each way but, as with the owls, we are rotating only on the horizontal plane, whereas flies can rotate their heads on both the horizontal as well as the vertical plane. And some of the greatest rotators are those that have to catch mobile prey, as with the robber flies, or those that are trying to catch hosts for their larvae e.g. the Tachinidae. Krapp is measuring how quickly a visual stimulus is acted on and then studying the neurological process of this to see how they are able to react so damn fast! Although there are not as many neurons in a fly's brain as either an owl or a human they are very good at multitasking, and it is this that Krapp is interested in for the further development of supercomputers.

A sibling family to the syrphids, the Pipunculidae, are one of the more amusing looking families of flies. They are commonly referred to as the big-headed flies because most of the 1,400 or so species have notably large heads in relation to their bodies. Stephen Marshall, a Canadian entomologist, in his wonderful book *Flies: The Natural History and Diversity of Diptera* (2012), describes them as looking like 'small syrphids (hover flies) with outlandishly enlarged heads'. And this is an apt description. Their fantastic heads appear to be entirely formed of their compound eyes, with three ocelli nestled neatly at the top and a pair of delicate antennae protruding from the front of the face (they do have mouthparts but they are often very simple). This family comprises three subfamilies: Chalarinae, Pipunculinae, and Nephrocerinae. Chalarinae have a semiglobose head (the backs of their heads are flattened), but Pipunculinae and Nephrocerinae have a completely globose head – that is totally spherical. Perched on tiny necks these appear to be some of the most delicate of flies, but their behaviour quashes this theory.

All the species in this family are parasitoids, and the larvae develop either in bugs such as leaf and plant hoppers, or in crane flies. Interestingly in these flies it is the females that have the enlarged ommatidia – that wonderfully named love spot that I talked about in male flies. The need to find a host for the larvae may be more challenging in this case than the need to make the larvae in the first place! She is often seen flying deep in the vegetation, looking for suitable hosts for her larvae.

In comparison to us humans, flies have limited colour vision. We have both indirect (genetic) and direct evidence of this, including an experiment that trained flies to prefer one colour and then seeing if

The heads of Pipunculidae are resplendent with ommatidia – looking very much like a microphone. The female has enlarged ommatidia at the front to help her find hosts for her offspring; the males have eyes that touch.

they could distinguish this from other colours (they did). Unlike the bees, which have been extensively studied, only a couple of species of fly, including the drone fly *Eristalis tenax* (Syrphidae), have been examined in this way. Colour is an important feature used by plants to attract pollinators, and we have known for a very long time that the reason plants evolved colourful flowers was to bolster their sex lives. A German naturalist, Christian Konrad Sprengel (1750–1816), a man made famous for his work on plant sexuality, was the first to realise that flowers are mostly for the insects, the pollinators. He understood that cross-pollination – the mixing of two or more genetically different parents instead of just self-fertilization – provided the fittest of offspring. The plants have 'realised' that in trying to get the insects to help them reproduce, they need to flower in colours that advertize themselves to the pollinators. A huge number of these pollinators are flies. More than half of the described families of flies feed from flowers and are thought to be pollinators.

Once again, the flies show a subtlety of skill that few of us appreciate. They don't just land on every colour they see, indiscriminately. They will weigh up the pros and cons of visiting a specific flower in terms of what they have learnt from previous trips to similar flowers and adjust their behaviour accordingly – the faster they can react to what is presented, the greater their chance of making the best choice – should they stay or should they go?

Flies are able to make comparisons between two types of colour – UV (which is picked up by R7) and blue or green (R8). But as we humans know colour is an odd thing – the blue and black or white and gold dress that caused an internet sensation in 2015 is a classic reminder of how we perceive colour differently. The same colour can vary in intensity, its spectral purity, its contrast etc, all of which will have a different appeal to the insects, and which is further affected by time of day and the weather. Flies vary in photoreceptor sensitivity, which along with their ability to smell (see Chapter 3) will affect what flowers they are attracted to. What looks to us like flies haphazardly choosing flowers is in fact a series of specific responses to a combined series of stimuli, both natural and learnt.

We can see this innate response with the drone fly, *Eristalis tenax*. Research has shown that freshly emerged specimens are initially attracted to yellow objects, i.e. those that humans see as yellow, hoping that these are flowers (a lot of flowers are yellow, so this makes sense). Over time, flies learn which ones provide them with the most nectar rewards and hone this basic instinct accordingly. I came across an amusing example in a *Dipterists Forum* bulletin, showing how this innate love of yellow for this species can have unexpected results. Member Christine Storey was sitting in her local coffee shop when she spotted a cake adorned with yellow marzipan flowers. Settled on the marzipan was an *Eristalis tenax*, and the poor fly seemed intent on trying to feed from them. Let's hope it at least gained some benefit from the sugars.

Flies see colour differently to humans and also detect light differently. We see unpolarized light – when we are outside on a sunny day it is very hard for us to determine where the light comes from as it appears to come from all directions. Our eyes can't see it any other way. But the eye of a fly can distinguish what direction the light is coming from, it filters all of the light, a process called polarization. This enables the animal to orientate itself by seeing waves in only one direction – they can work out the intensity and so the direction of the source. The marmalade hover fly *Episyrphus balteatus*, so-called because its colour resembles chunky marmalade, is a good example of a species of fly that not only is able to tell the direction of its flight but is also good at remembering directions i.e. it navigates. This species is found globally but some populations of these flies are static and some are from regions far away. The UK has two such populations, a resident year-round population, and one that migrates south to the warmer climes of the Mediterranean and down to North Africa in the colder months, returning to the UK once more when it is warm enough.

In males it is not just the vision in eyes that can differ. It is usually the males that are pursuing the opposite sex or defending territories and so need to be showing off their fitness or headbutting opponents. We have the antler flies, the stalk-eyed flies and the moose flies to name but a few of the many species that have eyes on stalks or strange protrusions on their heads.

We have already discussed the enormous amount of work undertaken on *Drosophila melanogaster* in relation to understanding the brain, eyes and so on, but to give all the credit in this family to just one of its species may be a little unfair as there are other truly extraordinary members among its 3,950 species described to date, many of which look amazing. A new species of Drosophilidae from the genus *Diathoneura* was described in 2003, by American entomologist Tam Nguyen, who was then affiliated to the American Museum of Natural History. From just two specimens caught in a malaise sample – a tent with no sides that passively collects insects – he set about describing this new species. What was odd about the males – and only males were found – was that there was a protruding appendage on each 'cheek'. Named *Diathoneura*

The drosophilid fly, named after the devil in Johann Wolfgang von Goethe's *Faust* *Diathoneura mephistocephala*, with its wonderfully cheeky protrusions.

mephistocephala, after the devil in Goethes *Faust,* and 'kephale' meaning head in Greek, it does very much look as if it has horns. A number of species from this family are yet to be described, especially in the tropics, so let's hope for some more bonkers looking species.

However, because this species has only been described from dead specimens, we can't be 100% sure about their purpose, but as with all things weird, it is probably coming down to the age-old activity of flirting. Luckily for us, there are five further families of flies that have similar protrusions, and so we can see if this theory holds true. One of the most intriguing are Richardiidae, a relatively small family of 175 species in the superfamily Tephritoidea, which also includes another six odd-looking families of flies. As with *Drosophila*, these have some amazing head features, including these 'antlers' that appear to be poking out of the fly's eyes.

These protrusions are called cheek processes and are not caused by a growth, as the name implies, but are formed from cuticle extensions of the cheek. So, what are they for? Well it's mostly due to sex, or rather the means to enable procreation. Males flirting with the females, by fighting other males to show their worth, have many territorial disputes over areas that contain the best sites to lay eggs. And what better weapon or deterrent than a face horn? In 1989, an amusingly titled scientific paper *The Horny Antics of the Antlered Flies* was published, in which American researcher Gary Dodson described the behaviour of another genus of these 'horny' flies, *Phytalmia*, in the family Tephritidae, the true fruit flies. Their preferred egg-laying sites are in the rotting wood of trees. The males show off at, fight at and guard these sites to ensure they can attract females.

When a male *Phytalmia* tries to steal another male's territory, to gain access to females, the defending male is not going to roll over easily. They become increasingly aggressive, eventually rising up on to their middle and hind legs so they are locked in combat, and a rather violent game of pat-a-cake ensues. Even in very closely-related species, the horns can vary considerably, suggesting rapid evolution. A paper published in 2017 by Dr Rosaly Ale-Rocha and her post doc Lisiane Wendt, then both at the Instituto Nacional de Pesquisas da Amazonia (INPA) in Brazil, described six new species of *Richardia*, a genus of antlered flies in the Richardiidae family, that all had different facial protrusions. These showed more diversity than other features in the adult flies, indicating that the evolutionary pressures for these to change was greater than for other secondary sexual characteristics – those not directly related to the actual act of copulation.

Many animals exhibit complex mating strategies – we are all familiar with the annual rutting of deer and the gesturing of peacocks but this also happens on a much smaller scale with flies. Stalk-eyed flies are a firm favourite for many a dipterist, these adorable little creatures with the craziest of heads – the hammerhead sharks of the air. As their name suggests, they have things sticking out of their face. But they are not horns; they are eyes, supported at the end of often quite long stalks.

The development of the optic nerve in these stalk-eyed flies is a fascinating process. When the adult emerges from its pupal case, they immediately start gulping air through their oral cavity (mouth), which is then pumped through ducts in their heads, and within 15 minutes they have completely inflated their stalks, a process made possible because the cuticle is still soft. This is merely the inflating stage. The growth of the stalk has already occurred, over a much longer period of time, inside the pupa. The development of the eye in the pupation phase initially appears to be similar across all flies. However, what happens next with this fly is when things start to get funky. The initially thick optic nerve narrows and gradually elongates, coiling around itself in the short eye stalk. The cuticle starts to look like a densely corrugated roofing panel, progressing into what appears to be a miniature accordion. Once the adult bursts out of the pupal case, it gulps in air which forces the eye stalks to rapidly expand with their already elongated neurological connection. If you ever find yourself, say, in an Ethiopian forest, and you come across a newly hatched stalk-eyed fly, settle yourself in as they are wonderful to watch – I spent hours just staring at these creatures while researching there. I also collected them in Costa Rica. Any reputation I had with my students as a proper grown-up scientist was lost when they saw me walking and giggling around a tree chasing the male fly.

The family Platystomatidae, the signal flies, contains the fly with the longest stalks. The males of the genus *Achius* are spectacular and the first specimen ever discovered of *Achius rothschildi* is one of the treasures of

The type specimen of *Achius rothschildi*, with its incredibly wide eyes – the largest of all hammerheads.

the Natural History Museum, London, named in honour of Sir Walter Rothschild. If we scaled this up to a human equivalent we would have eyes that spanned 9 m (29½ ft)!

But *Achius* is not the only genus of weird-headed individuals from a family of over 1,200 species. One of the most recognisable flies in Africa – and it's rare that I get to say that about one species of fly across an entire continent – is *Bromophila caffra*. It's a fairly large fly, true, but its coloration is what strikes you, a rather dark body and a very red head. It is commonly called the buzzard signal fly and although the species had already been described by Justin Pierre Marie Macquart, a brilliant French entomologist in 1846, no one has really furthered our knowledge since. American coleopterist (studier of beetles) Ted McRae proposed the common name only in 2009, in his blog 'Beetles in the Bush'. His reasons were that they, like their bird namesakes – and he means the 'buzzard' of North America, the turkey vulture, *Cathartes aura,* not the European buzzard, *Buteo buteo* – have conspicuous red heads in comparison to their bodies and eat 'repulsive foodstuffs', faeces. The stunning red head is thought to be a defence mechanism, 'Don't eat me, I am very unpleasant'.

The aptly named buzzard signal fly, *Bromophila caffra*.

However there has been no research on the reason behind this dramatic head colouring. From 1902, Guy Marshall, a British entomologist, worked on medically and agriculturally important insects in Africa, and conducted a series of investigations and experiments over five years to study insect mimicry and warning colours. He mentions the buzzard signal fly, describing it as a sluggish, conspicuous creature that 'ejects a yellow liquid from its mouth when handled'. Nothing has been published since about this fly or this 'substance' but Marshall noted that it is similar in appearance to other distasteful insects and that baboons won't eat them.

This bright head coloration can be seen in other species of flies. *Thyreophora cynophila*, once thought to be extinct but rediscovered in Spain in 2007, is a species belonging to the Piophilidae family. They are commonly called the bone-skippers, as they are rather partial to eating bones (as well as cheese), and the maggots are able to jump. The adults have dark black bodies and a bright orange head, or as Daniel Martin-Vega (the man of maggot fame in the previous chapter) describes in a 2010 paper a 'luminous head'. Information about this species is scarce, and what we have is at best described as slightly fanciful as it is based on a lot of hearsay. French entomologist, André Jean Baptiste Robineau-Desvoidy, in a publication from 1830, states that the male 'only looks for darkness and dried-up corpses. In the gloomy light of his phosphoric head, he throws himself upon the... bones'. But at the time, Robineau-Desvoidy stated that he had never 'had the good fortune to meet' this species and the information was from Amédée Louis Michel le Peletier, a well-known hymenopterist. No information was ever recorded from le Peletier himself.

The head and its features are such a complex region to deal with and I've only just scratched the surface. Research with diptera goes down two distinct paths: those that study *Drosophila* to understand complex processes such as neurological networks, visual recognition systems and so on, and those that are drawn in by the looks and natural history of all flies. Both groups recognise that although small, flies are ingenious, beautiful and highly capable creatures that far exceed those of other more lauded animals. Who knows what we could discover if there was more crossover between the two worlds?

CHAPTER 3

The antennae

*Smell that? You smell that? Napalm, son. Nothing else in the
world smells like that. I love the smell of napalm in the morning.*

Lieutenant Colonel Bill Kilgore, Apocalypse Now (1979)

THE ANTENNAE, along with the eyes, are the most obvious
sensory structures on a fly's head. For a long time they were solely
used to differentiate the nematocerous and Brachyceran flies until it
was shown that relying on them exclusively had little taxonomic value
– there are too many exceptions to the rules. Flies were originally
split into two subphyla – Nematocera and Brachycera. The name
Nematocera was coined by André Marie Constant Duméril, a French
zoologist (1774–1860), in 1805, nearly 50 years after Carl Linnaeus
published his hugely important taxonomic work on classification.
Although originally we have clumped all the flies with 'long' antennae
together, this is now considered a paraphyletic grouping, i.e. there is
no exclusive common ancestor to this group. Nematocerous means
'thread-horns' and most of the time that holds true but here's a case
where it does not. The non-biting midges, the chironomids, as already
discussed, are incredibly abundant around aquatic ecosystems. I mostly
dealt with their larvae earlier, but anyone who has run any terrestrial
traps in these habitats will testify to the abundance of the adults. So,

The plumose antennae of a male mosquito – its fluffy features are essential for
finding a female.

A female chironomid, which has short antennae, unlike the male which has feather duster antennae.

what was I looking at down a microscope that looked like an adult chironomid but had short antennae? Well, it transpired that it *was* a chironomid but a female and in their case the antennae are short. As with most female flies it is adorned with very few hairs and is very different to the feather duster antennae of the males. Males with very plumose antennae are common in many species of flies, as it is these structures that help the male decipher all the air-borne clues to enable him to meet the future mothers of his offspring.

Antennae are composed of three parts which, in keeping true to the nature of flies, vary massively across the order. The first component is the scape, which attaches the antennae to the head, and is the only part of the antennae to have its own muscles. Next to this is the stem, or pedicel, an important section for the fly as it contains the Johnston's organ, which I will cover shortly. These components are both fairly uniform in most species of flies. It is the third antennal section, the flagellum, that is much more variable and often a good diagnostic feature. It is composed of a number of segments called flagellomeres. In the more primitive diptera it can sometimes be exceptionally long, with many segments (as with the male chironomid) while in more advanced diptera, such as house flies, it is represented by just a single segment (with a derived appendage called an arista). The number of segments varies considerably not just across the order but also within families and genera, and between the sexes, as does the number and type of sensory receptors, also known as sensilla. The antenna is important in terms of hosting sensory equipment, including chemoreceptors (for detecting chemical changes), thermoreceptors (changes in temperature) and hygroreceptors (which detect moisture); these combine to provide an incredibly efficient gauging system. And what's more, adult flies also use their antennae to hear. I wonder what it would be like to hear with our nasal hairs?

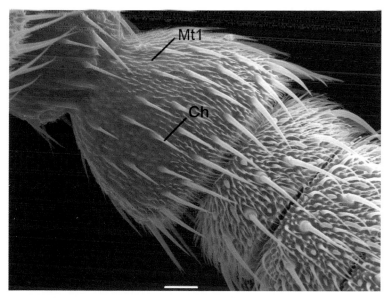

Pedicel of *Coboldia fuscipes* (Scatopsidae) antenna. (Ch-sensilla chaetica; Mt1-microtrichia.)

The numerous sensilla, resembling either bristles or pegs, are attached in pits to one or several nerve cells, which connect directly to the central nervous system. These structures are some of the first to detect the slightest change in conditions, such as the direction of the breeze or the presence of chemical cues from a nearby mate. Ernst Haeckel (1834–1919), the notable German biologist, philosopher, artist and physician, was the first to use the term sensilla to describe these structures, one of the many names he gave to science. He was a great 'labeller' – he gave us many words that we now use frequently, such as ecology, phylogeny (evolutionary history) and interestingly, also First World War.

One of my favourite descriptions of sensilla are those of an adult horse stomach bot fly which have been described as 'resembling rabbit ears' – indeed they do. The bunny ears of the clavate sensilla of *Gasterophilus nasalis*.

These sensilla vary enormously in shape and size depending on the role they play, and they may be described as clavate (club-like), chaetic (bristley), trichoid (hair-like), coeloconic (with pitted pegs), ampullac (with pegs in tubes) and basiconic (with pegs). That may seem a bit of overkill but the fly needs to obtain a lot of information, and fast. Its life may depend on it, and its sex life is utterly reliant on it. Even something as small as a fly requires a huge number of sensory devices to enable it to survive and reproduce – in relation to their size they may have to travel vast distances. All of the fly's different sensors combine to provide the most complete picture of the fly's immediate environment.

Irrespective of shape or size, all sensilla protect the delicate nerve endings that they are connected to. These are the communicating pathways, a series of neurons, to the insect's brain and central nervous system. These neurons are known as afferent neurons, and they convert the stimulus into a nervous impulse. Added to all these sensory structures are very small hairs that protrude directly from the cuticle called microtrichia, and these may form very dense mats. In *Drosophila* there are around 1,200 afferents that combine to form the antennal nerve that heads off to the brain, specifically the antennal lobe, which then sorts, filters and codes the information before sending it off to other parts of the brain, including the mushroom bodies. The flies are able to not only understand basic chemicals such as carbon dioxide but also patterns of smells – the combinations of odorants that combine to make an object unique – that reveal so much about it. The unique aftershave of a specific animal created by the pheromones it releases coupled with the cuticular hydrocarbons (CHC) – the waxy, smelly, substances that coat all flies – provide an assembly of smells, some of which will be more important than others and all of this needs to be filtered from the millions of other smells floating around.

The diversity in shape of the sensilla had historically led to confusion about their function as some structures may have more than one sensory function. A contact chemoreceptor, for example, may act as both a mechanical (physical) and chemical receptor, which are used when the fly comes into contact with a host or mate – so there's lots of groping and sniffing. And just because some sensilla may look similar to each other on the outside, it doesn't mean what is hidden inside will follow suit – internally there may be several different smaller structures, as seen in sensilla coeloconica, for instance.

A very small, but economically important species is *Sitodiplosis mosellana* (Cecidomyiidae) commonly called the orange wheat blossom midge – a long but descriptive common name. The larvae of this species feed on the developing grains in the wheat – they eat its ear – resulting in shrivelled grains, a decrease in germination, and further water and fungal damage thanks to them damaging the outside layer of the grain. And the results can be devasting. In Canada, the annual economic loses, considering not just yield reductions but also the cost of pesticides, can be over $100 million a year.

Because this species is such an economically significant pest, it has been studied in some detail. Both females and males use their antennal sensilla to locate hosts, the females for suitable oviposition (egg-laying) sites for their eggs, and the males as they know the females will be there. By understanding how they do this we can better develop tools and techniques to prevent them devastating the crops. In 2016, Wang and co-authors used both scanning electron microscopy (SEM), and transmission electron microscopy (TEM) to zoom right in on the different sensilla found on the midges' antennae. These microscopy techniques create images by sending a focused beam of electrons, either to scan the surface or penetrate through the specimen. Among all the different types of sensilla on the antennae there were numerous peg-shaped sensilla coeloconica, of two different types – imaginatively named Type 1 and Type 2. By comparing similar structures in other flies, the authors were able to hypothesize that the Type 1s were either olfactory (the chemoreception that forms the sense of smell) or humidity receptors, because of the presence of deep, longitudinal grooves and the many finger-like protrusions on the surface. These are very receptive to airborne chemicals. The Type 2s were thought to be contact chemoreceptors because, while these had no projections, they had a pore at the end of each sensilla. Understanding what sensilla are on the antennae and their function helps build a comprehensive model of, in this case, the olfactory system, and we can use this to manipulate the fly's behaviour by manipulating the plants odours.

As we learn more about the type of sensilla on a fly's antennae, we can also make assumptions about the behaviour traits and life histories of these insects. For example, anophelines (the subfamily Anophelinae of mosquitoes that includes species that transmit malaria) have large, grooved sensilla coeloconica whereas the culicines (subfamily Culicinae, the other half of the mosquitoes, which are vectors for West Nile virus and dengue fever), don't. Why the anophelines have the larger sensory structures for sensing humidity and smells than culicines is still not known but such variations don't just happen for no reason.

Anopheles gambiae is one of the most important mosquito vector species globally when it comes to transmitting malaria, but, as with *Anopheles barbirostris* complex I wrote about in the Introduction, it also sits within a 'complex' of seven morphologically-identical species. We first realised this 50 years ago whilst studying their behaviour. The original members comprised three considered freshwater, one thriving in mineral-rich water, and two salt-breeding species (more species have been added since). *Anopheles gambiae sensu stricto* – the original name-bearing species – was freshwater, while another member of the group, *A. merus* was a saltwater species. Although very difficult to morphologically separate, we have found one clear, albeit very subtle, difference, and that is with the antennal sensilla. There were many more of a type of chemo receptor called sensilla basiconica on the saltwater species. Researchers have not yet discovered the significance of this, but they at least help to differentiate different species. We have now identified other differences in the distribution of olfactory receptors across human- and non-human feeding species, and this knowledge will hopefully lead us to better identify species that are potentially problematic.

Another good example of how effective these sensilla are is with another species of mosquito, which I make a special point to collect when I am in the Caribbean, *Deinocerites cancer*. I love this species – it has adapted to live in an unusual environment – crab holes. The females of *Deinocerites cancer* don't feed on the blood of crabs, but on the birds in the surrounding coastal lowland wetlands. The males, as with all male mosquitoes, just consume nectar. Most of the time the males are found in the borrows, even after emergence. Males emerge from the pupal case earlier than the females, then sit on the water in the burrow, guarding the pupae, suspended beneath the surface, that are about to burst open – a process we call pupal guarding. The male has exceptionally long antennae packed with olfactory (for smelling) and mechanoreceptive sensilla – way more than is average in mosquitoes. Male mosquitoes usually have fluffy antennae for good hearing, but the antennae in *Deinocerites* males are essentially 'bald'. Instead these males have a greatly enlarged final flagellum segment, packed with very specialized sensilla coeloconica that are thermoreceptors – he finds and guards, what he hopes is a female, by monitoring her warmth, rather than listening to her, below the surface.

I, along with other fly botherers, know that the difference between one species and the next can be miniscule – the arrangement of bristles, the tilt of the antennae, and the shape of the genitalia. How on earth does the male check that out, especially if the female is flying past him at a rapid pace? We identify many species by their genitalia – a difficult characteristic to identify when the adults are alive as they are often internal. Looks alone don't help the males, so instead he relies on smell. In many species the females help the males in knowing that they are both suitable and ready by releasing a specific sex pheromone that starts the mating ritual off, causing the male to dance, wave its wing or strut about as if possessed, like that embarrassing uncle at a wedding.

As well as being tuned into each other's smell they can also listen to each other. And males listen very intently. At the base of a fly's antennae, in the pedicel or second antennal segment, sits the Johnston's organ, which is a group of yet more sense organs that detect motion in the flagellomeres. The Johnston's organ is named after American entomologist Christopher Johnston, who, in 1855, was the first to note its auditory function in mosquitoes, though it is found in all insects (but not the other Entognathous hexapods e.g. the springtails and bristletails). The number of cells within the organ varies across orders and in some species of flies. In mosquitoes and chironomids there is a very developed Johnston's organ but some species lack any other antennae sensilla. Many male mosquitoes have the most developed Johnston's organ of all insects that we know of. They contain 15,000 sensory cells called neurons – comparable to the human cochlea, the snail-shaped part of our inner ear that is the main player in our hearing ability, and home to 16,000 neurons. The plumose flagellum of both male midges and mosquitoes may resemble fluffy deeley-boppers, but they play an essential role in helping the Johnston's organ to detect the much quieter female.

The males being nectar feeders have no interest in humans – their sights are set on a more romantic prize and Johnston writes beautifully

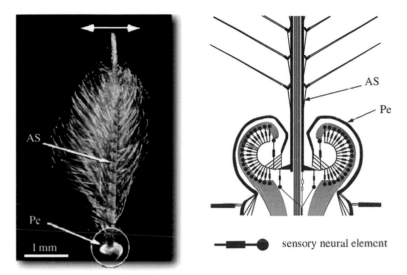

The remarkable hearing organ, the Johnston's organ, which is highly developed in the mosquitoes. (Pe, pedicel and As, antennal shaft).

on the habits of Culicidae males, describing males as 'timid creatures', who restrict themselves to damp and foul places. Unobserved, they sit and wait. He does not smell for her but instead listens for her distinct, whining wingbeat caused by the vibrations of her wings, known as a flight tone. It's the sound that many of us dread when trying to get to sleep at night. The sound waves caused by the movement of her wings pushes the feathery flagellum around, an action picked up by the Johnston's organ, resulting in a neuronal response – he hears her. He then hones in on her with astonishing accuracy, as we have learned from a study on the aptly named elephant mosquito (it's huge) *Toxorhynchites brevipalpis*. First of all, the male mosquitoes use their elaborate antennae to pick up the sound of a female flying past; he is able to amplify her signal to really listen in on her. If she gets very close, he can then tone down this amplification process so as to not overstimulate himself (none of us like people shouting in our ears). In 2006, Joseph Jackson and Daniel Roberts, both then University of Bristol researchers, found that the male mosquito responded to the sound of a passing female in a way that bore similarities to how humans responded – the cocktail party effect – so called as it refers to the ability that we have in being able to hone into one voice or being able to hear our name being muttered, even in a noisy room. Male mosquitoes would make great spies as they are able to increase their acoustic input a staggering 45,000 times. He mimics her wing beat frequency, creating the same but noisier sound, which combined with the quieter female, creates this greatly amplified sound. Matthew Su and co-researchers at the University College London, Ear Institute, found that those species with the largest swarms were able to turn up the volume to the highest – an ability that greatly helps males in a very busy and noisy environment.

The antennae of the female elephant mosquitoes are far less elaborate, with just a few hairs on each flagellomere. This is common not just in this species of mosquitoes but across the flies. Females don't listen so attentively to the males or even other females. For females, its mostly down to chemistry, but looks are also important. In 2015

The antennae of *Toxorhynchites brevipalpis,* a male that really does listen to females.

Floris van Breuge and co-authors based at California Institute of Technology and the University of Washington, published their work on the mosquito *Aedes aegypti*. The female first smells for the presence of carbon dioxide (as a plume of odour) – that tell-tale substance that animals, including us, exhale as a waste product – and this alerts her to the nearby presence of a food source. She then starts to visually search for the source of the smell – the females have more ommatidia than males to help this. Humans become visible to mosquitoes at a distance of 5–15 m (16½–49 ft) depending upon environmental conditions. Once she has located her victim she then uses local cues, such as body temperature and sweatiness, to help her pick the perfect landing site.

A family of small flies, Corethrellidae, is closely related to the mosquitoes, and like them, the females of the family need blood for egg development – but in this case they only feed on frogs' blood and we commonly call these frog midges. Much smaller than mosquitoes, they appear to have downsized to suit their hosts and only target a particular family of frog. Male frogs tend to be the noisy sex (though there is an exception – the females of the Emei music frog *Babina daunchina* protest in a series of clicks if the male's sexual performance is interrupted), and female frog midges take advantage of this by selecting just them as a source of blood. Unusually for flies, both the males and the females of this frog-biting family have very fluffy antennae and swollen pedicel, housing the very sensitive Johnston's organ. Although Corethrellidae have not been studied anywhere near as much as the mosquitoes (there is not quite the same imperative to do so), because of their relatedness and their structural similarities, we can make some confident assertions about how she locates her meal. We know that a mosquito's ability to detect sounds relates to wing length and beat frequency – the smaller the mosquito, the higher the note, or frequency. Researchers have caught specimens by playing a recording of male frogs to lure the blood-lusting females into a trap. They had assumed these midges would have a similar 'ear' to that of mosquitoes, ones capable of selecting frequencies of between 400 and 500Hz. However, this wasn't the case. It turns out the females are not just tuned in to one frequency, but many. They discovered that they are not as host-specific as the mosquitoes, and often feed on a variety of frog species, including the diverse types of tree frogs from the genus *Hyla*. Two North American entomologists, Art Borkent and Peter Belton, found that the female midges must have a hearing range of between 1000 Hz to 2900 Hz to correspond to the calls of the frogs.

These frog-biting species, like the mosquitoes, have incredibly long hairs on their antennae and it is believed these are essential for hearing. Borkent goes on to hypothesize that the ancestors of these species may originally have only used sound for finding mates and relied on chemical cues such as carbon dioxide for locating their frog hosts. With time she came to rely less on chemical but more on audible cues resulting in the development of much fluffier antennae.

Frog-biting midges have been around since the Cretaceous period (145 to 66 million years ago) and have been fine-tuning their antennae ever since. The variation in the antennae – the length and shape – has diversified greatly. The nematocerous flies, for instance, usually have between seven and 15 sections, or flagellomeres, with some featherlike, others threadlike and some pluriarticulate (many jointed). Although no longer an important phylogenetic feature, they are still important taxonomically and there are some stupendous antennae amongst these species.

Take the family Deuterophlebiidae, the mountain midges, who we met earlier in the maggot chapter. This is a tiny family of only eight species, but it's a very interesting family. Adult males emerge from the fast-flowing streams before the females, and die pretty soon after copulation, which may be just a few hours later – these are some of the shortest-lived adults in flies. Their short life span is unusual but not as unusual as their antennae. These males have a tiny body of between 2 to 4 mm long, but their antennae are up to 16 mm (0.6 in) long. The idea of me having ears that are four times the length of my body is a rather amusing one. These antennae are even more unusual in the fact that they are curled, almost spring-like in appearance, seen in no other insects apart from a group of extinct saw-flies (part of the Hymenoptera). Why should only these amongst the flies have such features? What is even more incredible is that they are rigid, with

An adult male *Deuterophlebia vernalis*, with its incredibly long antennae.

muscles only around the base! The specimens in the collection at the Natural History Museum, collected over 40 years ago, still hold their curly-wurly form.

Not only are the antennae of the male Deuterophlebiids odd, but the males can't walk, thanks to large, clawed pads at the end of each leg. What's with all this crazy apparatus? Well, these little wonders mate on the wing, not the ground, in large mating swarms. The males patrol for females over very fast-flowing streams. The weird feet are thought to prevent them from breaking the surface tension of the water, and so avoid drowning by accidently falling into the river. And the antennae are for defending and securing their place within these swarms – they demarcate their little aerial territories. The lengths that these male flies go to securing a female are very admirable.

Not to be outdone by the incredibly long antennae of the Deuterophlebiidae, are the Keroplatidae, one of the five families of fungus gnats. In Keroplatidae, there is a fairly large genus called *Macrocera*, of which 193 species have been described to date (four times as many as any other genus of Keroplatid). Like all fungus gnats, these flies are small and delicate, with generally humped thoraxes. What makes the *Macrocera* genus stand out is that many of the species have very long antennae, and the name itself 'macrocera' simply means 'big horns'. One of the star performers is the stupendous male *Macrocera phalerata* .

The *Macrocera* antennae are too long to be controlled by muscles, but instead are manipulated by the movement of haemolymph – an insect's equivalent to blood – which is pumped up and down the inside the antennae. The flexibility of their antennae is determined by the extent of the pumping, and also elastic membranes between each segment that holds them together and allows them to move as one long unit.

How they, and other insects, pump fluids around each antenna is quite something. I have already mentioned that flies have several so-called pulsatile organs – their 'hearts'. One of these extra auxiliary hearts is located in the fly's head (which means they don't have to worry about choosing between their head or their heart). As previously discussed, insects have an open circulatory system, where the haemolymph is pumped into an open cavity, the haemocoel, which bathes the

The very long antennae of *Macrocera phalerata*.

organs in nutrients and gases. However, this system only works well in small animals with a small metabolic requirement. The challenge with this system is to ensure the extremities are kept well supplied with nutrients, hormones, oxygen and so on, but also to ensure that the antennae don't get clogged up. These narrow antennae need to be both moved around and regularly purged of waste metabolic products; if the haemolymph does not flow quickly enough the fly will become groggy. The antennal heart operates separately from the main heart and functions independently, pulsing away at the base of the antennae, pumping haemolymph in and out. Not only are there antennae hearts, but also wing and leg hearts, abdominal hearts, and even some genital hearts in the females of some species, and the presence and size of these mini pumps varies across species.

In the Brachycerans, the chunky robust flies, the antennae are shorter, with (bar an exception that I will come too) at most eight segments. Many have the antennae tucked away at the front of their head, in an inverted dome called a cephalic depression. Why these flies have shorter antennae is not known, but it is probably due to a combination of natural (environmental) and sexual selection. The Brachyceran antennae have a greater concentration of sensilla in a smaller area, which may provide improved signal reception at a cheaper cost to the fly – nervous tissue is very expensive in terms of how much energy is needed to run it. The first two segments of the antennae are very similar to those of the nematocerous flies, with a scape and a pedicel, but changes have occurred with the flagellum- the number of flagellomeres are reduced and a stylus or an arista has developed, both of which are mechanoreceptors. The stylus is formed from the terminal flagellomeres (usually one or two but can be up to six) and is seen in the majority of lower Brachycerans. The fusion of the last three flagellomeres in nearly all Muscomorpha forms the arista. The first flagellomere in the Brachycerans is often larger than any succeeding flagellomeres, or there has been some degree of fusion of the basal flagellomeres resulting in a structure we call the postpedicel (or funiculus). Just to further complicate matters, the number of fused flagellomeres can even vary within a species. It's as if flies are hell-bent on being the most morphologically complicated creatures on the planet.

One such family whose antennae vary considerably is Stratiomyidae, Many of the species have spines on their thorax that resulted in the scientific name, from the Greek meaning 'soldier' and 'fly'. The Germans have a better common name, Waffenfliegen – the armed flies. In the UK, our common names for species have continued this military theme; there are majors and generals, along with legionnaires and centurions, and they are some of the most brightly coloured and distinctive flies. It is not just their body patterns that make them stand out but also the length, shape and structures on their antennae. It's almost as if this family was the testing ground for a whole variety of antennae. It's not just the change in length that varies across this family but the way in which the antennae are held. Some are poker straight while others are obviously elbowed. In other insects where these elbows have been studied e.g. social insects such as the ants, the 'elbows' enable them to use their antennae as 'arms' to touch and taste objects. These flies also appear to have arista, but these are not formed from the last three flagellomeres and as such are called arista-like stylus.

Getting the correct parts named is vital to ensure that we can correctly identify the species and figure out how it is related to other species. By being able to correctly name a body part we can compare or contrast a species with similar features seen on other flies. One of my many favourite males is within the Brachyceran family Mydidae, called *Perissocerus arabicus*. Honestly, it has one of the most ridiculous sets of antennae I have seen. These males are incredibly hirsute and, with chunky antler-like antennae, they look like

Perissocerus arabicus male and its chunky antler-like antennae.

miniature flying reindeer. The male is a lot smaller than the female, and whilst copulating, he can be observed being dangled upside down from the female – not the most dignified of mating positions. The female antennae, whilst bulbous, do not look like the pair of inflated antlers that adorn the male's head. The characteristic shape of this bulbous postpedicel helps us distinguish it from species in a closely related genus – *Rhopalia*. But this is of no concern to the fly – he is just concerned with finding a mate and winning her over, a common reason for such exaggerated features.

Many of the lower Brachyceran families, including the mydids and the soldier flies, have wonderful antennae. Another special example is the the genus *Rachicerus* in the family Xylophagidae, the awl flies. Containing around 60 species, this genus was previously considered to be its own separate family, partly due to the form of its antennae. The genus contains species where the antennae of both the males and females have about 20 flagellomeres that are pectinate – they look like they have massive combs coming off their heads. Comb-like antennae are relatively common in other groups of insects, such as the beetles, but are found within only a few groups of flies – and this is the only genus within the Brachycerans that have this shape. The added marvel is that both sexes have adopted this pectinate shape, hardly any female flies across the whole of Diptera have this form. And guess what? We have no idea why.

Weird antennae are normally the domain of the males and one of the best examples of male extremism is in the genus *Borgmeiermyia*, from the parasite family Tachinidae. Comprised of just four species, the genus was first described in 1935 by American dipterist Charles Townsend from an individual that turned up on the windowsill of Father Borgmeier's house in Rio de Janeiro, Brazil. Father Borgmeier's calling may have been the church but he, along with hundreds of other preachers, often had more than a passing interest in entomology. German born Borgmeier spent the majority of his life in Brazil and had more than a soft spot for phorids. Townsend was a known authority on Tachinidae at the time and out of respect to the Father, he named the new genus after him.

The males of all species in the genus *Borgmeiermyia* have some of the most incredible antennal features, because they are multibranched or 'multifissicorn'. They look superb! The adult males are relatively small, averaging 3 to7 mm long, but what they lack in size they make up for in beauty. Their antennae are so conspicuously weird, and there are at least three further genera in this family where the males have such sculptured antennae.

The actual part of the antennae that has exploded in form in these flies is the third antennal segment, and the arista arises from here. The third antennal segment, amongst its other roles, serves to detect any changes in wind speed or direction velocity – the very inverted bowl shape of the male's face and branched arms of the third antennal segment could further amplify any changes. The females of these species have massive sausage-shaped third antennal segments, a structure so large that it is presumably important for host location and sequestering – to date one host has been identified for this genus and it was a grasshopper,

Head of male *Borgmeiermyia* sp., showing elaborately branched antenna.

insects that are very good at fleeing. Male dipterans are known to locate the hosts as females will often be nearby.

Any mechanical changes in the antennae are registered by the Johnston's organ at the base and translated into a neurological response. All flies are able rotate their antennae to determine changes in wind direction and speed – very useful to hone in on your prey as well as stopping you getting blown off course. Researchers have found that during flight *Drosophila* antennae were positioned in the opposite direction to that of the fly's flightpath. This has inspired engineers to incorporate similar feedback mechanisms into minute flying machines to help with stabilization issues.

Sawyer Fuller and collaborators at Harvard University, were just such a group of bioinspired engineers, and have been working on what they call a microbotic 'bee'. I am putting bee in quotes here, as the robot only has one pair of wings, and so technically is a fly. They gave this little flying robot ocelli, simple light sensitive 'eyes', to help with navigation and maintain stability. To fly from A to B requires a whole lot of processing of external information and any response needs to be fed back to the wing, e.g. changes in wind speed, direction and so on. The Mars Rover was able to perform a similar set of stabilization instructions on exploring the uneven surface of the red planet, but it took the rover two minutes per calculation. That's just not practical for a fly let alone us larger animals – two minutes is a very long time and would result in many a tragic ending for many an individual, splatted against many a windscreen. For these little flying bots, the engineers thus set their sights on flies to investigate how these flying marvels were able to react so quickly to their environment. From these studies, they created copycat false antennae that could detect wind speeds down to 0.1 m/s and as such were very sensitive to changes and this, coupled with the ocelli, enabled the bots to fly in a controlled forward motion. This robo-insect is less than 100 g (3½ oz) and smaller than a UK ten pence piece (just shy of an inch). Being able to achieve stabilized flight in something so small is a wonder – thanks to the insects, and in particular, the flies.

Antennae are not just essential for stabilization but also for detection. One of the primary functions of the antennae is to help flies sniff out food. The fusion of the flagellomeres in the higher flies has not reduced their sensory capabilities, merely concentrated the apparatus. Most natural objects have a smell, whether we choose to accept this about ourselves or not. We humans are very smelly: we live in smelly places and we like smelly things. Nearly everything we do gives off a series of chemical cues, which either attract or repel other species. The flies have mastered their smelling capabilities, which is why they are often the first at the scene. Folks marvel (or I guess more often, despair) at how quickly a fly arrives at a piece of fresh meat left out on a kitchen surface. While writing this chapter, I found my thoughts focusing on an enormous calliphorid – a sparkling metallic bluebottle – that was circling my living room. Distracted from musing about antennal shapes to what this fly was doing there, I realised there must be trouble somewhere in my flat. Sure enough, there was a freshly decapitated grey squirrel in my study that my monster (cat) had left for me, one which this tenacious fly had smelt out before me, and I presume, had left some of her offspring on.

The higher flies have used these smelling skills to target an enormous feeding range, including many different flower types, and as such are incredibly important pollinators (the plants that pretend to be rotting corpses e.g. *Rafflesia* and the titan arum *Amorphophallus titanum* are my favourite examples of this). And this is a global phenomenon. No other group of pollinators has such a range from the coldest climes to the highest mountains. They flourish in these extreme regions, where many of the Hymenoptera, the other massively important pollinating group of insects, aren't found (Well, except the bumblebees with their fluffy coats.) Flies not only sniff for plants though; many are attracted to fungi, including the wonderful truffle-finding flies from the genus *Suillia* (Heleomyzidae). If you fancy some lovely black truffle for your soups over the winter months, you need to follow *Suillia pallida* around, and if you fancy some of the summer varieties of truffle to grate over your pasta, you need to track *Suillia tuberiperda*. Other flies have less refined tastes and can't get enough of faeces, including several families of dung fly, while others are attracted to the pungent aromas given off by corpses – be they fresh or many years old. And there is one family of flies that feed on the entire range of these food sources, and that's the Phoridae, or scuttle flies – arguably the most ecologically diverse family of animals on the planet.

One species of phorid, *Conicera tibialis*, is commonly referred to as a coffin fly. It is not unusual for it to feast on human corpses that have been dead for three to five years, with many generations having spent their entire lives underground on this one carcass. Researchers have even found this species in a buried human corpse exhumed after 18 years. This corpse had been laid at a depth of about 2 m (6½ ft), which, if these adult females (for it is them that dig down) had been human-

Conicera tibialis – the affectionately named coffin fly with its pear-drop shaped antennae.

sized is the equivalent of digging 3.2 km (2 miles) down. Who knows if this population had been living down there for years, or whether new individuals had been recolonizing the body, attracted in the first instance by plants at the surface.

Another genus in this family, *Pseudacteon*, like eating ants – at least, their parasitoid larvae do (as seen in Chapter 1). This may seem a rather gruesome lifestyle, but one of the species is turning out to be very useful to humans. For the last decade or so, the adults of the species *Pseudacteon tricuspis* have been released in the southern USA to control the imported fire ants – *Solenopsis* spp. The females find the ants by smell. A 2007 paper by Li Chen and Henry Fadamiro, both at Auburn University, examined the antennae of the flies in great detail, and found that both males and females have sensilla trichoidea, sensilla basiconica and sensilla coeloconica. The first two, the authors summarized, were for olfaction, and the latter was for thermo-hygroreceptory functions, such as the detection of CO_2, temperature and humidity. When combined, these provide a useful array of tools for seeking hosts. The males need to be able to locate the ants as well, as the females are mostly found near them. Just to make doubly sure that he finds and picks a suitable female, the male also has an extra set of sensilla trichoidea.

For more on sex and antennae, we turn to the Caribbean, a region known for its hospitality – good music, food and drink, warm weather, sultry beaches, glorious sunsets, and a destination for many a romantic honeymoon. Most of these qualities have not gone unnoticed by some of its smaller inhabitants. The Caribbean fruit fly, *Anastrepha suspensa* – a true fruit fly in the family Tephritidae – thrives there as an economic pest of soft tropical fruits. Its other common name is the guava fruit fly, a nod to its feeding preferences. The male, when in courtship mode, deploys two techniques to woo his intended. Not content with the traditional wafting of pheromones, the males have songs – calling songs and precopulatory songs – much like some of the islands' holidaying humans. The calling song, a pulse, is emitted in the 'lekking' stage, which is when the males congregate to parade and posture. In an atmosphere thick with pheromones, the virgin females listen with their antennae and react to these bursts of noise – unless it's a small male emitting them, sadly – in which case the virgins turn a blind antenna as his genes are not worthy of her attention. The male sings the precopulatory song to the female once he has mounted her, just before and during the early stages of copulation, presumably to keep her interest – it's always useful to keep the female entertained. Confucius wrote that music produces a kind of pleasure that human nature cannot do without – obviously many female flies can't do without it either.

Tephritids contain some of the most agriculturally impactful species, and with over 5,000 species across the globe, that impact is far-reaching. With larvae that are mostly phytophagous – plant munching – their cost in agricultural ecosystems can be enormous, and so pesticides are employed to control them. But we are all too aware of the effect pesticides have on the environment, thanks to the likes of the great American activist Rachel Carson and her book *Silent Spring*, published in 1962. However, this has not led to the cessation of pesticide use, rather just the development of different pesticides. These often fail on two counts – they don't eliminate the pest species, and instead they often eliminate the natural predators. A way to fool the pests to leave the fruit alone, or only deliver the pesticides when the pest is reaching critical levels, is a much more effective method of control. Since the 1950s we have started using integrated pest management approaches, in which we use many different methods, including biological as well as chemical, cultural and physical control methods, to prevent agricultural attacks rather than just relying on pesticides, with varying success. By understanding how these economically important species communicate with each other and tailoring our chemicals to be more species specific, we can develop methods to disrupt them without harming the wider environment – a product that we call a biorational pesticide that has a minimal impact on all species apart from the intended pest.

Take, for example, the oriental fruit fly, *Bactrocera dorsalis*. Originally from Southeast Asia, this little migrant has spread to many countries around the world and its distribution mirrors its cosmopolitan diet, which includes many cultivated and wild fruits. As with many fruit fly species that have been studied, the males of this destructive species

The males of the destructive oriental fruit fly, *Bactrocera dorsalis*, have hundreds more genes associated with odour receptors in their antennae than females. They detect chemicals from the fruit in the hope of finding a female there looking for a spot to lay her eggs as shown.

respond strongly to methyl eugenol. This is a naturally occurring chemical produced by more than 450 plants, which affects growth, structure, reproduction, and communication (yes, plants communicate with each other). The males respond strongly to this chemical, found in the fruits, leaves and oils of species such as apples, oranges and bananas, as they know that the females are looking for these fruits in which to lay their eggs, and so smell for the compound in the hope of chancing on a mate.

By determining the chemical composition of the odours, we can mimic these chemicals to make traps to control these species. And we can also think about how to make these species less attentive to these odours in the first place. Zhao Liu and colleagues from the Southwest University in China have tried to identify the genetic instructions for genes associated with odour perception. In their paper published in 2016, they looked at a strand of the ribonucleic acid (RNA), from male and female *Bactrocera dorsalis* flies and found that males had hundreds more genes associated with odorant receptors in the antennae than females – many of which are linked to pheromone detection in other species as well, such as *Drosophila*. As much as we are desperate to limit these species from getting to our food, they are desperate to eat too – their very survival depends on it.

The composition of odorant and gustatory (taste) receptors in both the antennae and palps of many fly species differ depending on the food source. For instance, the tsetse fly *Glossina morsitans* (Glossinidae), found in many African countries, feeds greedily on the blood of mammals. They can decimate whole populations of livestock by transmitting diseases such as nagana, which weakens the immune systems of the animals. Tsetse flies have far fewer taste odorant and gustatory receptors when compared to other diptera: they have 46 odorant receptors whereas *Drosophila melanogaster* has 68, and just 14 gustatory receptors in comparison to the much larger 73 found in *D. melanogaster*, and this is thought to be in response to their rather restricted diet.

Professor Dong Zhang and collaborators at Beijing Forestry University in China used several different microscopic methods to explore the receptors in the antennae of two species of lesser house fly, Fanniidae, in 2013. This is a very cosmopolitan family of flies, comprising 285 or so species with a particular fondness for humans, hence their common name. Not as familiar a name as their larger cousins, the Muscidae – house flies – but they are still of particular importance when it comes to decomposition and the socially important offshoot of this, forensic entomology. Zhang and the team used various techniques, including laser scanning microscopes, to create 3D structures from multiple-stacked images to reveal the tiniest structures. They found that the sensilla basiconica, thought responsible for detecting smell, comprised multi-chambered pits. These, they thought, must make the fly more efficient at trapping the odours of decomposing flesh than if the sensilla were just lying flat on the surface of the antennae, or in just a single-chambered pit. And there were more of these sensilla than previously thought. Earlier investigations with other flies had mainly used scanning electron microscope techniques, which weren't powerful enough to reveal these pit features

Antennae are not only pretty in their form, but amazingly diverse in their many functions. We have only just scratched the surface in trying to figure out what all the structures are for, and what they do, but we have already applied the knowledge we have in developing artificial structures to help us explore our planet. *Drosophila melanogaster* were the first animals in outer space (from this planet at least), launched on a rocket in 1947 and, miraculously, recovered alive (the rocket had a parachute). Who knows, this species could once again help us with space exploration, by inspiring the next generation of probes and robots with advanced mechanical and chemical sensors.

CHAPTER 4

Mouthparts

To lose a lover or even a husband or two during the course of
one's life can be vexing. But to lose one's teeth is a catastrophe.

A Little Night Music, Hugh Wheeler, 1974.

THE MOUTHPARTS in flies are as spectacular and as diverse as their choice in food. And not just between the species, but also across the life stages. All larvae have mouthparts, though some can be very simple in structure, and are adapted for chewing or sucking at foods that range from plants to flesh. One such flesh-feeding group is the genus *Philornis* in Muscidae (the house fly family), which are found in the Americas, from Florida down to Argentina. All of the 50 or so species in the genus have larvae that are dependent on birds for their nutrition; to date they are known to parasitize over 150 birds, from hawks to hummingbirds. Different species have larvae that either feed on birds' faeces or on the blood of nestlings and occasionally of adults, sometimes attaching themselves to the outside and sometimes burrowing through the bird's skin and feeding on blood and tissue. The adults are not so invasive in their feeding habitats and are all free-living, feeding on decaying matter. One of these species has got itself into a lot of hot water, as it is a parasite on a group of rather famous birds. Twenty years ago, *Philornis downsi* turned up on the Galapagos Archipelago and started feasting on Darwin's finches.

The mouthparts of the stable fly, *Stomoxys calcitrans* – the horny piercing apparatus.

That may be a bad story for the birds but a good example of the separation of diet between the larval and adult stage. The mouth parts of adults are very different to the larval structures as they are adapted for sucking liquid food, a very beneficial situation. The hemimetabolous species, as well as some of the holometabolous species such as the honey bees, have the same diet throughout their lives. Adult honey bees forage for nectar and pollen upon which they feed themselves and the rest of the colony, including the immature bees. Any change in the quantity and quality of this food source impacts across all generations. This is very uncommon for flies, where the adults and larvae often live in very different environments and feed on very different food. And to feed on all these different sources, many modifications have happened.

The heads of flies (and other insects) have developed from the fusing of body segments, with all their associated appendages, over time. In a process called serial homology, original appendages have also become customized for various different functions. Some of these structures have become legs, while others have been modified to become the mandibles, maxillae and a labium – the main mouthparts for all insects. The mandibles and maxillae function in a similar fashion to our teeth – grasping, cutting and chewing the food – and have mostly been lost or greatly reduced in many species of flies, an obvious exception being the bloodsuckers and some other 'novel' feeders. Adult flies don't 'bite' down on you and munch away though; oh, they can rip you apart, shred you, cut your head off (ok, only if you are very small, say a couple of millimetres), but munching on you is not a feasible option. That is left to the beetles and grasshoppers to name but two orders, both of which have bitten me on numerous occasions.

Instead, many adult flies have strongly developed the labium into two distinct sucking forms. Mouthparts developed for suctorial feeding are called haustellate and, in flies, the mouthparts have gone one of two ways: either that of a flexible probing needle, a form referred to as stylate, as seen (and felt) with mosquitoes, or a flabby, spongy pad-like mouth termed labellate, as seen in the house flies. But as with all things fly, they have taken the basic plan and run with it, modifying various associated appendages along the way. The labial palps seen dangling down on the faces of grasshoppers are not seen as separate structures in flies (they

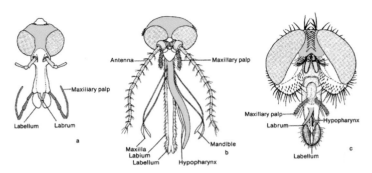

Mouthparts of various adult Diptera – a, nematocerous type (Tipulidae, *Ctenacroscelis*); b, piercing-sucking type (Culicidae, *Culex*); c, sponging type (Muscidae, *Musca*).

have different dangling bits) but have become fused to form the labellum – the lips. The term labellate in higher flies is in reference to the vast enlargement of this structure in comparison to the other flies.

Together with the flies, other suctorial feeding adult insects include the net-winged insects (Neuroptera) and scorpion flies (Mecoptera). These evolved 60 million years before the explosion of flowering plants in the Early Cretaceous Period but all contained members that had long mouthparts, called a proboscis. Many early insects were feeding on the 'pollination drops' from the gymnosperm plants, for example conifers, which don't have fruits or flowers – but there was something special about these insects and their long, probing mouthparts.

Archocyrtus kovalevi, an extinct genus in the family Acroceridae, was alive in the time of the dinosaurs and had the longest proboscis in relation to body size of any insect at the time, almost twice the length – back off *T. rex,* the most extreme animal was a fly! Alexander Khramov and Elena Lukashevich, both at the Borissiak Paleontological Institute of Russian Academy of Sciences, published a paper in 2018, on a single fossil specimen of *Archocyrtus kovalevi.*

Looking at the fossil, it wasn't immediately obvious that it even had a long proboscis. The long tube-like structure was not attached to the head, but lay extremely close, and some authors assumed that instead

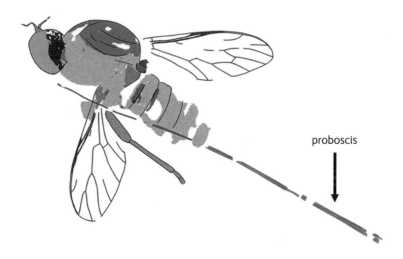

proboscis

Diagram of *Archocyrtus kovalevi* with its long tube-like structure, now thought to be its incredibly long mouthparts.

of being a mouthpart it was a piece of vegetation. Modern methods in microscopic analysis have shown this not to be the case, and it is indeed a tube-like structure. Khramov and Lukashevich theorized that *Archocyrtus kovalevi* was a pollinator of the plant *Williamsoniella karataviensis*, whose cones were found with this fossil fly. This cone did not have flowers, but it did have flower-like reproductive parts. And so, a long-tongue would have come in very handy to reach inside. The flies are some of the most ancient pollinators of the angiosperms, the flowering plants and, as such, played a very important part in the relationships between these plants and insects. As the angiosperms began to dominate, other species with tongue-like mouthparts, the bees and so on, developed siphon-like mouthparts, as seen in butterflies and moths. Flies were ahead of the game though, mutually evolving some of the longest mouthparts for some of the longest tubed plants.

Because of the lack of information on flies' life histories – many species have been described from individuals that have been caught in traps – we have limited information, and sometimes no idea, as to what they feed on. One genus with very long, almost mosquito-like mouthparts is called *Elephantomyia* in the Limoniidae family (the short-palped crane flies), and I collect these often in my malaise samples from Dominica. Within this genus is another example of an extinct species found in Middle Eocene Baltic amber with exceedingly long mouthparts, *Elephantomyia longirostris*. This proboscis was as long as the rest of its body combined – ranging from 2.66 to 4.20 mm (0.105 to 0.165 in). There are more than 135 living species of this distinctive genus, but even so, we know little about their feeding habits, some species have been observed feeding on nectar, but whether they are fussy about their diets, we don't know.

Or what about the Vermileonidae, the wormlions? This is a family of lower Brachyceran flies, of only 80 species, which superficially resemble crane flies. The family is known more because of its larvae, which dig out pits to trap and kill passing invertebrates who fall into them. Adults have rarely been collected or observed in the field; most of the specimens that we have in museums have been reared from larvae. But many genera in the family have species with very elongated mouthparts and a few have been observed feeding on nectar.

Lampromyia sp., an example of Vermileonidae with very odd proboscis.

Another group of flies with very long mouthparts, about which we do know quite a bit, are adults of the Nemestrinidae family, or tangle-veined flies. I rave about this family because some of the species within it have some of the longest mouthparts in relation to body size of all insects. The family uses these almost hose-like proboscises to suck up its favourite food – nectar.

But the world record for proboscis length goes to a rather famous hawk moth from Madagascar, *Xanthopan morganii*, commonly called Morgan's sphinx moth, though many know it as Darwin's hawkmoth. In 1862, Darwin received an orchid from Madagascar with an exceptionally long spur (the tube-like structure) (they can be up to 35 cm (13¾ in). This is the structure insects must penetrate with their proboscises. Darwin himself remarked: 'Good Heavens … what insect can suck it?' In his book *Fertilisation of Orchids*, published in 1862, Darwin predicted that a moth must be the pollinator, as he had studied many other moth-pollinating orchids 'the pollinia would not be withdrawn until some huge moth, with a wonderfully long proboscis, tried to drain the last drop'. In 1867, Alfred Russell Wallace agreed as, having come across *Xanthopan morganii* in his travels in continental Africa, had predicted that if it existed in Madagascar, that this species would be the pollinator. His prediction came true.

The proboscis of this moth is impressive, but a Nemestrinid proboscis beats it in relation to body size. *Moegistorhynchus longirostris* has a long name for a long mouth. Being able to fly around when you are just 2 cm (¾in) long, but with a mouthpart that does not coil and is over 6 cm (2¼in) long, is an amazing biological feat, in my humble opinion. A paper published in 2010 by David Barraclough and Rob Slotow, both at the University of KwaZulu-Natal, measured mouthparts of this species from localities across South Africa and recorded lengths up to 83 mm (3¼ in).

These beasts are an important part of the pollinating group of long proboscid flies found in the Cape region of South Africa, whose flora is especially interesting as it contains many species only found there – species we call endemics. The Fynbos, a belt of shrubland in this region, has an exceptional high level of diversity and endemism, including many species of *Erica,* a genus of plants from the Ericaceae, or Heath family. I was named Erica in honour of the clan badge of the McAlisters; this was a small sprig of heather worn on the bonnet or sash to show his or her clan, and so my parents (being both sciencey and scholarly) named me thus. Back to the Fynbos and its plants, of which *Erica* is the largest genus found in the region (and the rest of the Cape). A wide variety of insects are important pollinators of *Erica* species, and many have co-evolved with this plant. The short urn-shaped species are pollinated by bees, but the more exciting long-tubed species are pollinated by the long-proboscid flies.

The nemestrinids and other long-tongued flies, including horse flies, are prolific feeders, with enormous and rigid (non-coiling) mouthparts, begging the question: how do they cope flying around? Moths and butterflies, unlike the flies, are able to coil up their mouthparts – their proboscises are not formed of the more rigid labium but have developed from one of the maxillary appendages called the galeae – two c-shaped fibres that are fused together in the adult to form a feeding tube equipped with muscles and a blood and oxygen supply. So, what do the flies do?

Prosoeca sp. in the resting position. The proboscis lies beneath the thorax and abdomen, resting between the coxae.

Surely, they can't hold them out in front like a mounted knight jousting? Well they don't. A 2012 paper of a similarly long-tongued fly, that of the genus *Prosoeca*, another Nemestrinidae, revealed what happens to the mouthparts. Florian Karolyi and colleagues at the University of Vienna, Austria, and the South African National Biodiversity Institute, found that at rest, this fly tucks its mouthparts along the underside of the body, between its legs, projecting back beyond the abdomen (as we have seen with the Russian fossil, p.118). When needed for feeding, the long proboscis is brought forward and held out in front of the head. The change in angle between these two positions is up to 100°. And it does this by having developed a concertina-like feature at the base of its labium, a hinge it can use to rock its tongue structure back and forth. Moths might look good curling their mouthparts, but for me, having the ability to tuck such a huge mouthpart under their bodies is even more ingenious.

The evolution of feeding tubes is known widely among the pollinators, but perhaps the most notorious 'suckers' are the ones that feed on blood, the sanguivorous species. The most well-known are the mosquitoes – the females to be precise as they need the blood meal for egg development – although not all adult females are bloodsuckers. Those that are, are among the most harmful 'vectors' – that is, carriers – of pathogens of humans and other animals, and so understandably have been exceptionally well studied. The first thing the female needs to do is find her bloody food source. She has to detect a 'feeding station', a process determined by chemical, visual and temperature cues. The smells are picked up by the palps, those long sensory appendages around the mouth. The smells that send these mosquitoes and other biting insects into a frenzy are carbon dioxide, and an alcohol called octenol, or mushroom alcohol (one of the many organisms that produce it naturally). We all know that we breathe out carbon dioxide but we also emit clouds of octenol in our breath and sweat – what a pleasant thought. Interestingly, it has also been found that mosquitoes seem to like the smell of wine, so maybe leave a glass at the side of your bed to lure the mosquitoes to that rather than yourself (though you would have also have to stop breathing as well, because of the CO_2 issue...). Once she is nearby, she starts to pay attention to the subtler compounds that the host gives off, such as carboxylic acids (a wide range of acids), which determine exactly where she will land and ultimately strike.

Upon landing on her host, she uses her proboscis to detect the very small, thin-walled blood vessels called capillaries that criss-cross our bodies just below the skin. The exact mechanism for this has only recently been determined. In 2015, Je Won Jung and a team of researchers from Seoul National University, published their findings about how mosquitoes used their proboscis to locate blood vessels and feed without being detected, because, as they wrote, 'unsuccessful probing might alert the host animal to their presence, which may result in considerable risks'. I would consider death one of those 'considerable risks'.

The team discovered sensory hairs on the proboscis, containing olfactory receptors AaOr8 and AaOr49 (terribly catchy names). These receptors are activated by chemicals found in the host's bloodstream, enabling the female mosquito to locate the blood accurately and rapidly. To show that AsOr8 and AaOr49 were critical to this, the team switched off the genes that expressed these and instead of taking 30 seconds to

The multi-faceted mouthparts of a mosquito – nature's finest needle.

locate and then feed, it took the females 15 minutes. At that speed even the slowest of us may notice the intrusion and the days of that individual would be numbered.

The piercing proboscis of a mosquito (and many other flies) is not a single structure, but instead comprises six separate elongated mouthparts called stylets held together when not in use. These lie alongside each other to form the fascicle, or syntrophium, encased in the thickened outside cover, the labium. The labium is the part that bunches up on the surface, as the mosquito pierces through the skin, and is what protects the delicate apparatus when not in use, as well as providing physical support when they are.

To locate a suitable place to penetrate her host, the female mosquito vibrates her stylets, a bit like starting up a drill. The region at the end of the maxillae, referred to as the laciniae, have tiny teeth on their edges and act as drill bits, tunneling through the skin and then acting as anchors to keep the stylets in place once feeding commences. Inside the fly's head are protractor and retractor muscles, pulling and pushing at the base of the maxillae. A fused elastic structure attached to these muscles enables a rapid penetrating movement. Once in, grasping structures, which are greatly modified mandibles, hold the host's tissues apart while the largest of the stylets, the labrum, starts probing around. It's on the end of this stylet that receptors AsOr8 and AaOr49 are located. Once blood has been detected, the labrum pierces the vessel and sucks up the blood, much like us drinking a Bloody Mary through a straw. To keep the blood flow going, she releases saliva from the hypopharynx, which contains a tranquilizer to numb sensation. She probes and flexes her stylets under the skin to find these blood vessels, like the arms of an octopus reaching into the dark to carry out a dangerous deed.

These amazing piercing mouthparts have inspired clever folks to create similar structures that can pierce our skin without the dreaded pain. In April 2018, Dev Guera from Ohio State University and his international collaborators from the Indian Institute of Technology, published a paper on how we can use mosquito mouthparts for our own benefit. They found that mosquitoes use an insertion force of 10–20 μN. This is apparently three times lower than the insertion force of an artificial needle. It was the vibrating effect of the fascicle that enabled the much gentler force, which damages the tissues less and thus doesn't produce a large nervous response in the host. Once more an example of brains over brawn. And they discovered that the firmness of the labrum, the entire structure housing the stylets, changed with frequency – the higher the frequency the stiffer it becomes. This means that it can be flexible or firm depending on what the situation requires. These bioinspired microneedles are not only smaller and less painful but, instead of being made out of an inflexible steel, can be made out of polymers such as plastic or resin. These new needles can be also used in MRI scanners, whereas metal ones cannot. They can be transparent for use in lasers or fibre-optics that require light, and they can have more than one point – so that two reactive components can be delivered precisely. I love the idea that the species most humans hate are providing the inspiration for safe and painless ways to help us survive.

Once all the piercing is done, the mosquito's head also has structures to help it suck up as much blood as possible. There are two suction pumps in the female's head, the smaller of the two at the end of her proboscis, and the larger one at the base of the throat, or oesophagus. To initiate feeding, the first pump opens to lower the pressure and draw the blood up into the proboscis. The second pump then does the same, causing the blood to be further drawn into the body.

Many female Diptera are both blood and nectar feeders, and so different techniques are needed to suck these two food sources up. A lovely example is with the subfamily Pangoniinae – a group of horse flies (Tabanidae) containing the large and very attractive genus *Philoliche*. It contains more than 130 species, all remarkable for their very long, non-retractable mouthparts. It is only the females that are blood feeders, but they also supplement their diet with nectar. Nectar feeders generally differ from the blood feeders in having soft-tipped rather than piercing mouthparts. These blood-and-nectar feeding hybrids have both!

We can effectively think about them having two kinds of mouthparts. As with mosquitoes, all but the labium part is involved with blood feeding and perform in the same fashion. But in these dual-feeding species, this is paired with a foldable, nectar feeding part, composed

Philoliche sp. collected by Torsten Dikow from the top of vegetated sand dunes near Port Alfred, South Africa.

of a massively extended labium, at the end of which is the labellum, articulated by a hardened part called a kappa – the distal part of the proboscis. The surface of the labellum is covered with sensilla basiconica, both short-socket and peg-shaped sensory cones to assist in sniffing out nectar. The female also uses the sensilla basiconica to locate blood in her host, and once found, she plunges her piercing mouthparts in! What is even more amazing about this genus is that these relatively short mouthparts, used for blood feeding, are tough and long enough to penetrate the skin of a rhino. This family of flies have subsequently gone slightly wild with their choice of blood sources and have been recorded feeding from snakes, crocodiles and terrestrial turtles as well as dining out on delicate flowers! I personally have been attacked by them, but with my ninja sweeping skills I swept them into my collecting net, and they now reside in the museum's collection.

A female *Forcipomyia* sp. (Ceratopogonidae) engorging.

The ever-expanding abdomen of a ceratopogonid. The fly rapidly grows as it feeds on its host.

One of my favourite engorgers of blood is also one of the smallest – the Ceratopogonidae or the biting midges; a group of flies that not many people like. These species are smaller than 2 mm (0.079 in) with mouthparts between 0.1 and 0.2 mm (0.0039 and 0.0079 in). Small, but as anyone who has suffered the 'Scottish Scourge' will testify, because of their fondness for hanging around in massive swarms, their collective impact is huge. Rates of between 2,000 to 3,000 bites per hour have been known. However, as only 0.1 mL or one-ten-millionth of a litre, is taken per bite, we would need 56 million feeding events at one time to drain a human. The mouthparts of the biting midge are perfectly adapted for penetrating their food sources – the mandibles have teeth on the rearward-directed side (the side facing the fly as it retracts it back to the body) and the muscles pull in this phase resulting in a cutting of flesh. Their abdomens are also vital as they are able to expand rapidly to cope with luncheon (see Chapter 8). These tiny midges feast on a whole host of animals, from the very large to the very small (including members that specialize in feeding on the opposite sex during copulation). We have examples in the collection that are attached to the wings of other insects such as dragonflies and stick insects. There is a wonderful blog by Israeli entomologist and photographer Gil Wizen showing a series of images of a biting midge feeding off a stick insect. As it feeds, the midge gets bigger and bigger like a balloon. Once it has digested its meal, its body will shrink back to its original shape. I am slightly jealous of an animal that is able to expand and contract like that without having to go to the gym.

For blood feeders to feed without their meal knowing about it is a tricky one, but it is also essential that predacious flies are not damaged by their meal. And the sanguivores and predators have developed a weapon to help them. It's all very well catching a grasshopper twice your size – as the top predators, the robber flies, love doing – but then to have to deal with their long limbs flailing around while you're trying to eat is tricky. Or, what about the poor robber fly trying to catch a stinging Hymenoptera, or a tiger beetle, or an assassin bug – all of which are formidable killing machines. Robber flies, and many other families including the horse flies, have devised a way to paralyze their prey, or parts of it, before settling down to eat from or on them. For many species of flies are venomous, and instead of administering it from a modified egg-laying tube, as do the bees, wasps and ants, they inject this venom into their prey in a way akin to the spiders.

That some flies are able to catch and almost instantly paralyze their prey by means of venom has been known for more than 150 years. One illuminating study was carried out in 1925 by Sarel Whitefield, a researcher at Imperial College, London. He was reviewing what was then known about the predator behaviour of blood feeders, as well as conducting his own investigations. He compared the effects of the bites of robber flies on grasshoppers, wasps and other flies, with similar attacks in which he used needles on his insect victims as a comparison (his descriptions of sticking a pin in a wasp's head and wiggling it around are very detailed). Needless to say, neither of these two procedures generally worked out well for the prey; but with an important difference. The fly bites resulted in rapid paralysis, followed by death in most cases, while the wiggling needles either had no effect, or resulted in a slow, lingering death, presumably due to brain damage. Whitefield's paper lists the reactions and observations of robber flies attacking prey in the wild. It reads like the notebook of a serial killer, 'died instantaneously'; 'struggled violently for a short time, then suddenly expired'.

Investigation into dipteran venoms has been very slow to progress, which seems a shame as we now know they are rather unusual. Ten new venoms have been described from robber flies, and six of these venoms include proteins that no other venomous creatures, bees, wasps, ants, snakes, conches, platypuses and so on, have. Unimaginatively, these 10 venoms have now been named Asilidin 1 to 10 (after the family name Asilidae). These venoms are purely to paralyze, none have enzymes that help digest food externally, but are instead toxic smoothies composed mostly of peptides (short chains of amino acids) and larger proteins, resulting in the pharmacological breakdown of the recipient tissues. But although this may seem terrible for the intended victim, these novel peptides are now being tested for use on humans to help in the treatment of pain, cardiovascular diseases and diabetes, to name but a few research avenues.

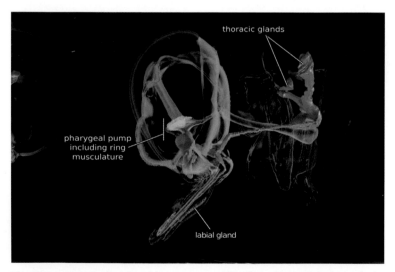

The 3D reconstructed venom delivery system of female and male *Dasypogon diadema*. The green parts are the ducts which transfer the venom to the tip of the proboscis.

These venoms develop in a pair of sac-like glands in the first couple of segments of the thorax, not in the head cavity but actually in the front of the thorax. From these two glands come two separate ducts that fuse together just before entering the head and then run parallel with the food duct before they enter, by a one-way valve, into a salivary pump and through the mouthparts to the intended victim. Whereas the mosquitoes and most of the other lower diptera produce saliva in the food canal, in robber flies that task is performed by the labrum, the structure that encases the proboscis.

Robber flies are maybe some of the most formidable predators, but I know of only a handful of stories where they have hurt humans, and most of those involve the victim trying to grab the fly, and the fly, understandably, getting angry. As we've seen, however, some flies are more likely to attack us as they typically feed on the blood of large mammals. This includes the aforementioned horse flies, but there are also offenders in the Muscidae family. Within the subfamily Muscinae, the smallish tribe Stomoxyini comprises only 10 genera and approximately 50 species, including the genus *Stomoxys*, commonly called stable flies. If you have ever been stabbed by *Stomoxys*, you can give testament to the fact that it's an incredibly painful event. The proboscis of *Stomoxys* resembles a very small bowling pin with sharp teeth at the top end, which they use to pierce through the skin of large mammals such as horses, hence the popular name stable fly. The hide of a horse is much thicker than a human's (and a zebra's, see p. 76) and so stable flies need formidable mouthparts to break through. I hear about these creatures whenever I lecture at Harper Adams University, set in the pleasant rural Shropshire countryside in the UK. This university has many agricultural courses and has its own working farm. And herein lies the problem affecting many of the staff, as this farm provides ample food for a large population of *Stomoxys* (its name means sharp mouth). I love visiting this university for many reasons, one of which is to see these flies as their mouthparts are something special. All Stomoxyini have this conspicuous 'horny' proboscis that is adapted for slicing and sucking the blood of many a mammal. What is fascinating though, is that although this blood extracting tool is large

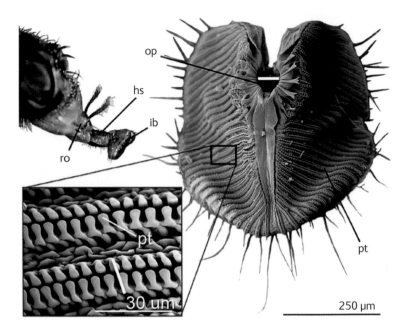

Spongy mouthparts can be divided into three regions: the rostrum (ro), the haustellum (hs) and the labellum (lb) (top left). The distal part of the labellum (SEM image, top right) has a network of pseudotracheal canals (pt) that connect to an oral opening (op) at the centre where the food canal is located.

and armoured, the fly has been recorded feeding undisturbed for up to 15 minutes – it would be interesting to see if they too are venomous and so numbing the hides of these large mammals.

The rest of the Muscidae though, have lovely fleshy pads for mopping up fluids – what we refer to as spongy mouthparts. For example, house flies often have large and bulbous structures, very obvious to the naked eye but many may have discrete mouthparts, as seen in many of the hover flies, kept tucked out of way till they are needed. Many of the species that have spongy mouthparts have what is called a rostrum, or beak, at the end of which, or contained within, are the mouthparts. In hover flies, these beaks can be very pronounced, as seen in the large *Volucella* species, while with other species the mouthparts can be elongated, as with *Rhingia* species. So-called sponge-feeding flies have an enlarged labellum that contains many grooves known as pseudotracheae, the channels through which the food passes. These are critical for 'persuading' the fluid to enter the mouthparts.

We have already talked about head pumps in mosquitoes, and they are not uncommon in diptera. They play a vital role in shunting food round the body, but the pumps alone are not enough to pull the fluids out through the food pores, such as the nectaries of plants. This is where capillary pressure comes in – the ability of fluids to move along very narrow spaces regardless of gravity. This action is happens in the very small pores of the plant, and in the food channels of the labellum in nectar-feeding species. The channels on the pseudotrachea create conditions suitable for the development of liquid bridges, drawing the nectar up out of the plant.

Eristalis tenax covered in pollen.

Pollinating flies don't just feed on nectar, they feed on pollen too. We know this because pollen grains still retain their shape when they pass through the gut, even though they are quite small at 0.015 to 0.2 mm. *Eristalis tenax,* one of the species of hover flies that we refer to as drone flies, is a very important pollinator. Some pollen feeders stuff their mouthparts between the anthers – the pollen containing part of the male fertilizing organ, the stamen, of a plant – and get the pollen that way; but not *Eristalis tenax.* It is a hairy beast, and pollen grains get stuck all over it and in the grooves of bristles. These hairs are vital for trapping pollen, which the fly later cleans off itself, consuming the pollen as it goes. Imagine being able to just roll around in your food and then consume at leisure.

Eristalis tenax is regarded as a mimic of the honey-bee, its appearance making it look more dangerous than it is to predators. But this may be doing an injustice to the hover fly. New Zealand entomologist Beverley Holloway believes that the similarities between honey bees and drone flies are the result of convergent evolution – that is, they look alike because they have similar food-gathering tactics and so have developed similar structures. Both have branched hairs and bristles with grooves for extra pollen storage and both leg scrape during hovering to transfer pollen to their mouths (or baskets in bees).

House flies have some of the more impressive fleshy mouthparts, barring *Stomoxys* of course. They use saliva to loosen up their food, for example, externally digesting some of the tough flesh into a suckable fluid. Some of these lovely house flies, along with blow flies and flesh flies, have prestomal teeth in addition to pseudotracheae. *Sarcophaga dux*, a medically important flesh fly that is both a forensic detective and an instigator of myiasis, has a series of particularly formidable looking teeth.

Although many adults have some impressive mouthparts, for the majority of species of fly, the main feeding stage is the larva. Not all adults feed or they have a limited diet often requiring just a small nectar or water boost. The exceptionally rare family Ctenostylidae – they have no common name – occurs across the tropics and is presumed not to feed as an adult because, although it has mouthparts, they

The prestomal teeth, located on the prestomal ridge that lies along the edge of the pseudotrachea of *Sarcophaga carnaria*

The most unusual of heads – branched arista and vestigial mouthparts of the *Lochmostylia borgmeieri* in the family Ctenostylidae.

don't function. With less than 20 species described, there have been problems in understanding their taxonomic position, and the family has been shuffled around the diptera tree. Taxonomists are happy that they are acalyptrates, but that's where it ends. Of the few Ctenostylidae species described, all but one species are known from fewer than 10 adult specimens. They are night fliers, which may explain the lack of specimens collected, and have no ocelli. Valery Korneyev, from the National Academy of Sciences of Ukraine, and one of the few experts on this family, speculates that with limited sight and non-functioning mouthparts the adult flies live but a short time. The females have been observed giving birth to larvae, not eggs, which although not unknown in diptera is not common within the acalyptrates. There is an assumption that these larvae are parasitoids, but we don't know for sure. Generally speaking, adult flies that have parasitoid larvae range from heavy feeders to not feeding at all, so we can't infer anything about them from this either.

Minute mouthparts are also seen in the Oestridae family, many of which have larvae that are parasitic. Take the fabulous bot flies within this

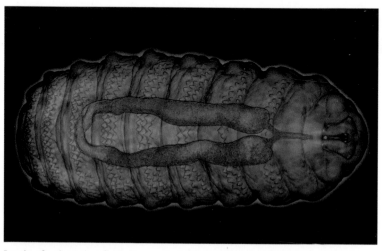

Inside of a sheep bot fly *Oestrus ovis*. A 3D scan showing the massive salivary glands.

family. Originally thought of as four separate families, this now combined powerhouse contains many adorably cute adults, but quite stomach-churning – actually stomach-churning – larvae. Living in the stomachs – but also in the nose and under the skin – of many types of mammals, the larvae of many of these species thrive, taking up to a year to develop, feasting on the fluids and soft tissues of their hosts. Entomologist Daniel Martin-Vega, who I have mentioned in previous chapters, has imaged bot flies using micro-CT, and has found that the larvae have huge salivary glands in their mouths, which can take up as much as two-thirds of the body's mass, especially in the sheep bot fly *Oestrus ovis*.

And this makes sense because, as Martin-Vega puts it, they are basically eating for two (themselves and their future selves). In the first instance they are producing saliva to release digestive enzymes, but they are also producing it to release chemicals to counter block the host's immunological response. And the larva will feast away, ensuring that by the time the larvae are adult, when these highly productive mouths have withered into vestigial mouths, they have enough reserves to fully function as an adult. Martin-Vega was able to compare adults of this species and *Calliphora vicina* (Calliphoridae), a species that does feed as adults, and there was a massive difference in the gut size of the adults. The adult sheep bot flies live at most a month, during which they must disperse and reproduce. But in other related species this may be less than a week. Either way, the larvae need to be efficient eaters to sustain the adult phase.

But just because an adult may not feed, does not mean it doesn't drink. A paper published by Californian authors Paul Catts and Richard Garcia back in 1963 describes how another bot fly genus drinks. The paper sounds more like an essay on the gin bars of Victorian London, *Drinking by Adult Cephenemyia (Diptera: Oestridae)*. These most hirsute of adults have, hidden behind their lavish beards, very small, spongy mouthparts, practically invisible to the human eye. They observed these adults drinking, so although they don't have the mouthparts to feed, they still have a mouth that is able to drink.

So many forms and so many structures mean some feed on flesh, while others on nectar, blood, fungi and even ant vomit in the case of some mosquitoes and some members of the family Milichiidae. All flies in this family are very small, 1 to 3 mm long, which is good because many have a sneaky habit of riding around on quite formidable predators such as spiders, robber flies and ants, feeding off the juices that leak out of the prey that these beasts are feeding on. This tactic has earned them the name the freeloader flies. Myrmecophily – ant-loving behaviour – is very common in this family, including the species *Milichia patrizii*. The name-bearing specimen for this species, the holotype, is deposited at the Natural History Museum, London, and it is also pinned alongside

Milichia patrizzi feeding from the mouth of *Crematogaster castanea*.

a *Crematogaster* ant. These ants lay scent trails with instructions where to get food and it is on these that *Milichia patrizii* ambushes her victim. If successful in isolating an ant, they grab onto the ant's antennae with their own, and the ant responds by stopping and crouching down. The fly then extends its mouthparts to the ant's which causes the ant to regurgitate its reserves. Alex Wild, the photographer who was able to record this interaction, stated that in one timed interaction, the fly fed for 10 seconds, while the ant was docile and motionless.

The ability of flies to feed from and on such a wide range of hosts and products has enabled this order to survive and flourish in the most inhospitable of environments or extreme of food stuffs. We should be grateful that flies are consumers of waste as well as pollinators – a life where I am swimming around in a quagmire of faeces or not having any black pepper, onions or carrots for my evening meals would definitely not be as optimal as the one I have at the moment.

CHAPTER 5

The thorax

One thing you'll say for skeletons, they'll always give you a smile.

Steve Aylett

IF THE HEAD is where much of the major sensory apparatus is located, then the next section along, the thorax, is all about using this information to enable the fly to move around. And to move, you need the right equipment. You need some decent muscles and then one tough outer layer, in this case an exoskeleton, to support them. The majority of animals, flies included, have their skeletons on the outside (no cupboards for them – their skeletons are on display for all to see), which need to be both flexible and rigid. The exoskeletons need to support the fly but in addition, offer protection, prevent desiccation and enable absorption and excretion.

The exoskeleton in a fly is known as the cuticle, which extends over the entire body and provides a good protective cover for the sensitive internal organs. While the cuticle of the larval stage is very flexible, to allow the characteristic pulsating 'wiggle' of the maggots and their rapid growth, in the adult phase it becomes much more rigid. In the bit between the larva and the adult, the degree of protection differs, with the Muscomorpha having a puparium. This puparium is formed

The thorax of the fly is where the muscles for locomotion are found, with some species having an the extra hump to fit them in (e.g. Hybotidae).

from the old cuticle of the larva that has become very tough and rigid, having undergone a period of hardening, regulated by the hormone bursicon. The pupal case has very low water content and the structure of the cuticle proteins has made it very strong; the larval cuticle in comparison has a higher water content enabling it to be very flexible. Inside the puparium, the fly needs its newly developing cuticle to be very malleable, so it can stretch and deform. It hardens after emergence, a wondrous process that I talked about at the start of this book. Adult life would be very difficult without this protective but flexible layer.

The process of emerging from the pupa is called eclosion, and the freshly emerged adult is referred to as a teneral adult, a term derived from the Latin 'tener' meaning soft, delicate and tender. This is a dangerous phase of the insect's life because, although mobile, many can't fly as their wings have yet to expand and harden, and their cuticle has not hardened. They are therefore much more vulnerable to predation and infection. Newly emerged tsetse flies experience what is termed the teneral phenomenon where, at this stage of their life, they are extremely susceptible to trypanosome infection, a group of parasites that causes sleeping sickness in humans. A tsetse fly may take several days to mature and they tend to lurk around rather than try to feed. It is when they take their first blood meal that problems start for these poor creatures. These young feeders have not yet developed a protective sheath, called the peritrophic membrane (PM), in their guts. This PM forms a barrier composed of primarily chitin and protein and, once formed, offers protection against invasion. Without the barrier, parasites can penetrate through the fly and some will end up in the salivary glands – from here they are just a bite away from infecting humans and other animals. Cuticle hardening, both external and internal, is essential in maintaining the health of the fly.

Although the cuticle is hard in the adult, it still needs to be capable of being manipulated to enable the fly to walk, run and perform its pièce de la résistance – its extreme flying skills. The cuticle is composed of two main layers: a thin outer epicuticle covered with cuticular hydrocarbons (CHCs) – the waxy odourless chemicals produced by the fly's genes and its lifestyle, and the larger inner procuticle, which contains the chitin.

These hydrocarbons form an effective barrier against desiccation. The blend of CHCs is distinctive to each species and can further change with age and reproductive ability, and so are unique to each individual.

The adult stage is often the shortest part of a fly's lifecycle and, with just two jobs – to disperse and copulate, they had better be ready for it. The legs, wings and associated muscles, and many of the gyrosensory organs, are all located on the thorax. All adult flies have three pairs of legs, but not all flies have wings (and I will leave that to the next chapter). You would think that the thorax would be pretty standard in shape across the diptera, but once more the basic body plan has been manipulated, and a large variety of shapes and sizes can be seen across the various families.

Flies are some of the most able creatures to have taken to the skies, and their wings are powered by incredibly strong muscles. Packed into their thoraxes, the sheer volume of these tissues has resulted in some fabulous humps. Some species, such as those found in the subfamily Tachydromiinae (Hybotidae) are very active, and have notably humped thoraxes, which we don't know the contents of but are thought to be jam-packed with muscles. The whole family belong to a group called dance flies, a name given to them because of their groovy flying activity, though this subfamily is more specifically called the fast-running flies because they restrict their moves to the surfaces of leaves, branches and stones. The majority of Tachydromiinae run around rather than fly, ambushing small prey by jumping on them and holding them secure with their grasping fore- and mid-legs, before they pierce them with their sharp mouthparts. They're highly mobile little assassins that need a lot of energy to maintain their slaying activities. Species in the bee fly family also have humped individuals but the flies that take the absolute biscuit when it comes to humped thoraxes are the small-headed flies from the subfamily Philopotinae, in the family Acroceridae – the hunch-backed flies. These flies are not assassins but pollinators.

The thorax is composed of three main sections and, in order, from the head segment along the body, are the prothorax, mesothorax and metathorax. Each segment has one pair of legs – indeed, it is effectively the fusion of three ancestral segments. One of the most interesting structures in one species is a novel development on the prothorax – ears. *Ormia ochracea* (Tachinidae) is a smallish species at around 7 mm (0.3 in) in length, found in the southern USA and down through Mexico, that can often be observed flying at dusk trying to find the crickets that are the hosts for their larvae. The light is not good at this time of day and so vision is not as useful. Instead, adults have developed 'ears', or rather a pair of sound-sensitive membranes, to track down a suitable host for their unborn offspring. The adult uses a rather clever mechanism. In a large animal, if a sound comes from the right there is a time difference between then and the sound reaching the left ear, and this helps the animal to pinpoint the direction from which the noise is coming. But in a fly, well, the distance between the ears is just too small. The distance between the two ears spans just 1.5 mm, and the arrival time of the cricket chirps to the ears is just one or two microseconds

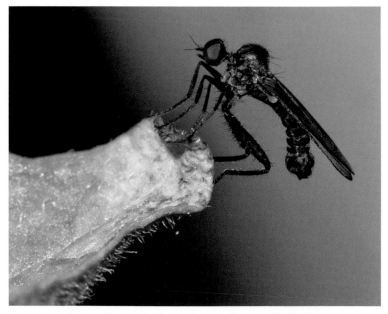

The amazing humped thorax (and well, the rest of it as well) of a *Hybos* sp. (Hybotinae, Hybotidae).

different. For this fly to determine where the prey is located by acoustic methods would be impossible – the neurons don't fire quickly enough for it to process and react to the noise. The poor tachnids would be sent mad as the crickets make a bit of a racket and would sound as if they were everywhere.

Instead, the adult *O. ochracea* flies have developed a way to slow down and turn up the volume of the sound waves they receive. The two 'ears' are basically sound-sensitive or tympanic membranes pivoted around a fulcrum, or base, which deforms the cuticle as it moves; it rocks like a see-saw. If the noise hits the ears at exactly the same time, there is no movement; but if there is any difference between the time each ear receives the sound this results in the ears being rocked due to a change in pressure. This system is able to amplify the sound as well as increasing the interaural time difference – the time difference between

The hunchbacks of the fly world, *Toxophora fasciculata* (Bombyliidae).

the sounds that arrive at the two ears. The ear that first hears the sound reacts vigorously in comparison to the other one, and the change in pressure between the two ears is calculated, at incredible speeds by the fly's brain, and in doing so the fly is able to rapidly pinpoint the origin of the sound. This creature has such acute directional hearing that it rivals the species thought to have the best directional hearing – we humans. Human ears are about 15 cm (6 in) apart and it takes us around 10 ms to make the same calculation that this species does in 50 ns. James Windmill and colleagues at the University of Strathclyde are one of the groups working on adapting this and developing incredibly small but highly directional hearing aids. At the time of writing, they had developed one that was just 3.2 by 1.7 mm (0.1 by 0.07 in) in size – they still have some work to go to get their mini-microphones down to the size of the fly's.

Many of the structures that insects use to stabilize themselves while flying are located on the thorax. These 'organs of equilibrium', as one of the greats of insect physiologists, British entomologist Vincent Wigglesworth, called them, are proprioceptors, which respond to movement and position. It is your proprioceptors that enable you to touch your nose when you are blindfolded. How a fly keeps its head level while the rest of the body manoeuvres is a fascinating process that relies on many different sensory systems: on the head these are the compound eyes, the ocelli and antennae, and on the thorax there are the halteres and legs (discussed in subsequent chapters). But there are further structures on the front part of the thorax that help in movement called the prosternal organs (POs). I have watched a video of a tethered hover fly that was able to rotate its head 270° when the body was twisted by the researcher, keeping its head level with the ground. The prosternal organs are a pair of matching patches of hair situated on a small protrusion above the first pair of legs, on the 'neck' of the thorax, and are stimulated when the head moves. The arrangements of these ever-so-tiny hairs vary across the different families of flies and have been located in all of the Brachycerans but not the nematocerous flies. All of the sensory subsystems work at different speeds but together provide a rapid stabilization system.

Nearly all these hairs and bristles (spines) have sensory functions. Collectively they are known as setae, but the terms 'hair' and 'bristle' have been used in a mish-mash fashion for so long that it's hard to figure out what people are referring to at times. Put simply, hairs are small, and bristles are large. Irrespective of size, sensory hairs or bristles sit in a pit in the cuticle and have a series of cells associated with them (one that produces them, one that produces the cuticle and so on). The most important for the purpose of their role is as a nerve cell that translates and communicates stimuli. These nerve cells are surrounded by glial cells, cells that offer support, protection and environmental stability (these cells will often envelope the nerves forming a protective cover). These glial cells basically keep things ticking along nicely whatever the situation, while the neurons rapidly pass the information to the central processing centre. The signals are sent from all over the body, e.g. in *Drosophila*, individual flies may have up to 6,000 hairs and bristles across the body. In all flies the smaller hairs produce a very rapid response, whereas the response is much slower with the larger bristles, and this response also varies according to direction of the stimulus.

The prosternal organs are not fully understood but we do know that they play an important role in enabling the head and thorax to coordinate. A huge roll of the head during a turbulent flight may result in an unworkable alignment of the thorax – the POs work by communicating to the brain both head roll and pitch. We have tethered flies in wind

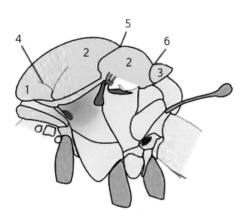

The sections of the thorax on a crane fly
– the head would be on the left if that helps!
1: prescutum; 2: scutum; 3: scutellum;
4: prescutal suture; 5: transverse suture;
6: scutoscutellar suture.

tunnels to see how their body angles change with air speed. The faster the air speed, the more angled the flies hold their bodies – very similar to the monocycle or gyrocycle in the film *Men in Black 3* with its stabilizing mechanism.

We've already identified the main three regions of the thorax, but these can be further divided to help us understand where specific features, such as hairs, are. The upper surface across the three regions is called the tergum further divided into three sections called the prescutum, scutum and scutellum. The side of the thorax is called the pleuron and the underside is referred to as the sternum. The entire structure is covered in hardened plates called sclerites, and there are different types depending on where they are located on the thorax – tergites (top), sternites (bottom) and pleurites (sides). These plates form the chest armour of the fly and provide structures for the internal muscles to attach to. The fly's thorax is essentially a box of muscles, but a very clever box of muscles. Because of the way the muscles work and how the thorax resonates, these tiny creatures are able to sustain flight. The thorax is also very important for ground locomotion, and it's essential, therefore, to keep it in good working order.

If this body armour is damaged the fly can heal it, in a rather similar process to our own. First, clotting of the fly's blood-like haemolymph occurs and then the underlying epidermal cells migrate into the damaged zone, secreting a new cuticle. The damaged sclerotized parts don't completely heal, but instead form a scar. This makes life fantastically difficult for a taxonomist as the resulting scars may hide a diagnostic feature. Identifying their many small and complex features is already difficult, but we also have to consider changes that might have occurred during an individual's lifetime. Scars might mean an absence of hairs where we might otherwise expect to see them, for example. Bit of a nightmare all round as these hairs and bristles are important taxonomic aids. German dipterist Ernst August Girschner, was the first to formalize the discipline of identifying flies by the arrangement of hairs or bristles on the body of the adults, making it an indispensable taxonomic aid, especially in the Calyptratae, which includes the Muscidae and Tachinidae – both families packed with bristly flies.

But it was Osten-Sacken who in 1884 coined the term chaetotaxy to describe using the arrangement of bristles to group flies. Translating the arrangement of these tiny little hairs probably results in more swearing by Dipterists than anything else (apart from sub-costal breaks on the wings, but that's a story for the next chapter). The hairs fall out – you are left with nothing but an empty pit – how on earth are we meant to identify whether the hair was pointing forwards or backwards now?

Before coming up with the term chaetotaxy, Osten-Sacken spent much of his time working with crane flies, horse flies and hover flies – none of which are particularly bristly. But his attention slowly turned towards the hairiest of hairy families, the Tachinidae. The distribution of hairs may seem quite random in some species, but as you look closer, you (hopefully) begin to see order. In the calyptrate families, which include the Tachinidae, the large bristles, called the macrochaetae, appear in specific patterns, which differ from one species to the next. In contrast, in acalyptrate families such as Drosophilidae, the same bristle arrangements are seen across the genus *Drosophila* and have been this way for well over 50 million years, when this genus first appeared, we describe these morphological characters as highly conserved unlike the incredibly variable sexually selected characters that change in a blink of an evolutionary eye.

From the second thorax segment the wings arise and these, due to their importance, get their own chapter (along with the halteres which are on the third segment). Both the scutellum and postnotum (= mediotergite) are part of the mesothorax, but the former appears to be part of the final segment. The 'scutellum', Latin for small shield, for indeed it looks like one, is often a very bulbous feature. The scutellum may also have very large protrusions or spikes, along with the obligatory hairs and/or bristles, coming out the back of it. A good spikey group are the Stratiomyidae, the soldier flies. I talked about their name coming from their spiny appearance on p.102 and again, the German name Waffenfliegen – the armed fliers – seems more appropriate to these beasts. These scutellar spines are so distinctive and diagnostic that British dipterists Alan Stubbs and Martin Drake, use them as the first feature for identification in their identification key to British

Beris vallata (female) from Sutton Park, Birmingham, UK.

Stratiomyidae. In the UK, the number of spines on their scutellum ranges from zero to six, in e.g. *Beris vallata*, and they are so easy to spot that even I can't fail to identify them.

Soldier flies are not the only flies with spines on their scutellum; other good examples are seen in the family Coenomyiidae, in the genus *Coenomyia*, which have a pair of often quite chunky spines; as do the stalk-eyed flies in the family Diopsidae. Spines on the legs and the abdomen make sense in terms of defense, predation or copulation, but why here? The genus *Teleopsis* (Diopsidae), the fabulous stalk-eyed beasts, have rather long spikes on their scutellum. A wonderful image taken by entomologist Ji Tan of a species in Malaysia shows the male, while mounting the female, holding on to her spikes, so maybe they are used to aid copulation for other species? But, as with many structures, they may have many functions and we have yet to determine them.

This is not the only peculiarity of the scutellum. There is a whole family of small flies that have developed something very unusual with theirs. The Celyphidae are a relatively small family of about 115 species. Predominantly they are found in the rainforests of Africa and Asia, though they have also turned up in South America. Regardless of where they are found, all have grown their scutellum over their abdomen, resulting in the most beetle-like of appearances, and earning them the

The love handles of *Teleopsis* sp. (Diopsidae).

A *Spaniocelphyus* from Singapore - oh what a pretty beetle, oh wait – are those halteres?

common name of beetle flies. Their scientific name comes from the Greek 'kelyphos', meaning pod or shell, and I dare anyone to look at these and not admire their beauty. OK those beetle-fans – the coleopterists – will argue that it is because they look like beetles, but to me they resemble engorged ticks with a snug cape.

Beetles are, for the majority, easily recognised by their elytra – a protective shell that covers their thorax and abdomen, formed by the hardening of their first pair of wings. These are held at rest over the second, membranous pair of wings that are used for flying. When most beetles fly, the elytra open, the hardened wings are held apart, and the second membranous pair of wings are able to move unimpeded. This is not the case with the Celyphidae as this abdominal covering has developed from the single scutellum. Although we are still uncertain as to why this feature developed, Richard Frey, a Finnish entomologist, in 1941, proposed two options. He suggested that the abdomen covering was either to improve buoyancy in flight, presumably by trapping pockets of air (although he didn't add anything further to this point and no one has commented on this since), or for ornamental reasons, i.e. to aid flirting.

Flies are some of the most ornate animals out there, with their eye stalks, cheeks with antlers coming out of them, and leg paddles to name but a few weapons in their arsenal of seducing accoutrements – and so to assume that Celyphidae use these elaborate shields for flirting seems perfectly feasible. The question of why they have such an enlarged scutellum has not been answered, and very few folks have observed these flies.

Celyphidae are not the only flies to have this scutellum development but they are the most extreme case. Two other families have also developed enlarged scutellum – the Chloropidae (frit flies) and Ephydridae (shore flies). This seems like a case of convergent evolution but, as with many dipterological questions, it remains a mystery for the time being.

Not all flies even have a scutellum, especially the wingless ones. The Braulidae, the bee louse flies, are all wingless, and more closely resemble hairy little teddy bears than anything else. They have no need for wings, riding around on the backs of bees, and subsequently their thorax is much reduced, and the scutellum has disappeared altogether. I will talk more about the position of flight muscles in the next chapter, but the muscle attachments within the scutellum are vital in keeping both wings coordinated with each other but, if you don't have wings, there's no need for this structure.

Tachinids are hairy all over but have some incredibly long hairs on their scutellum. The species *Phorocera obscura* is a prime example of

A very bristly male *Phorocera obscura.*

this – it is an incredibly hirsute individual with bristles extending back from the thorax to over half its abdomen. This species, as with all the *Phorocera* species, lay their eggs in caterpillars, but the tachinids have broad tastes when it comes to hosts, with many hosts not yet known. Although the majority attack caterpillars, tachinids will attack other arthropods, mainly insects but also scorpions and centipedes. The large and small bristles on the thorax have not been studied to a great extent with flies – and even when we have they have mostly been with the little *Drosophila*, but research has so far shown that many of these bristles are less sensitive to wind but more sensitive to being touched and groomed.

Another common feature of tachinids is an enlarged subscutellum, the bit underneath the scutellum. Not all tachinids have this feature and there are species in at least two other families that have this as well. The Mesembrinellidae, a family of just 36 species once thought to be a subfamily in Calliphoridae but now in its own family, and a new member of the Oestroidea superfamily, includes the genus *Mesembrinella*. All in this genus have a slightly swollen subscutellum and are just the most stunning of hairy flies. These flies superficially look like tachinids or hairy Calliphorids but have a very different lifecycle, and in fact are not closely related to either. They feed on decomposing animal matter and fermented fruits, but unlike many of the other tachinids these species nurture their larvae internally. They don't need to seek out hosts for their young. But they are incredibly restricted in their choice of habitat, and have only been found in humid, pristine forests. Recent studies have found a direct relationship between their presence and how untouched their environment is – the slightest mucking around in the forest by us and they are gone. We have not yet understood why these squashed-bottomflies are so fussy, or why their thorax bristles are so long.

The final segment of the thorax, the metathorax, is greatly reduced, and the metanotum is but a narrow sclerite between the mesonotum and the base of the abdomen. It is on this section that the very important

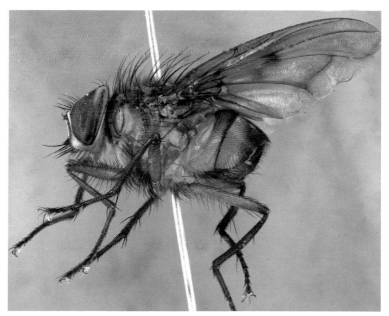

A male *Mesembrinella patriciae*, another stunning and very hairy fly with a slightly swollen subcutellum.

halteres are located and also a pair of thoracic spiracles – an all-important opening for air exchange which, coupled with the anterior spiracle, enable the fly to breathe.

The muscles dominate the inside of the thorax, but for these muscles to operate optimally they need a plentiful supply of oxygen. Can you imagine what would happen to the poor adult fly if it suddenly developed cramp in one of its thorax muscles? It would plummet from the heavens. Flies are supplied with enough oxygen by using the same system of racheae as used by the larvae – the formation of tracheae has resulted from invaginations (ingrowths) of the cuticle that creates a network of breathing tubes throughout the body, directly supplying oxygen to the tissues. Unlike the larval stage, very little respiration occurs across the cuticle (it's a lot thicker). Instead adult flies have two large air holes – called spiracles – down each side of the thorax as well as some smaller abdonimal ones. The thoracic spiracles are named the anterior or mesothoracic, and the posterior or metathoracic, spiracles, and they open up to supply oxygen to the respiratory system. As these air tubes penetrate the body they get smaller and smaller in diameter, ending in the tiniest of tracheae, the tracheoles, which are less than a micrometer (0.001 mm) in diameter, which themselves terminate in either open or closed tubes that are 200 nm – that is just 0.0002 mm – or simply put, rather small. These tubes branch all over the body including up through the wings, round the abdomen, through the legs and in the head. But these spiracles cannot be left open or the majority of species of fly would just dry out. Instead flies are able to open and close them. This process allows a uni-directional air flow – air flows in and then air flows out. This allows oxygen into the body and then the waste product, carbon dioxide, out. Also, very important for this are the hairs that surround the spiracle, which act like filters. Research has shown that the larval conditions of the individual can also result in a difference in the hairiness of the adult spiracles. Wadaka Mamai and colleagues, in 2016, published their research on the differences in spiracles from very closely related species of mosquitoes (those in the *Anopheles gambiae* complex), that are naturally found in different environmental conditions. When they reared these different species, under both dry and wet season conditions, they found that the hairs were thicker in all three species in the wet conditions than from those in dry chambers.

Each spiracle is set in a peritreme, a thickened cuticular plate that surrounds the opening. It is this feature that resembles the 'eyes' of the anal spiracles of dipteran larvae. As we've seen, spiracles are often protected by hairs to stop foreign particles entering and damaging the internal breathing tubes. In the blow fly *Calliphora vicina*, the front spiracle is obscured by a dense layer of dark orange pubescent-like hair that arises from the peritreme. Nothing solid is getting through this mesh to damage the breathing hole beneath, which opens and closes to regulate air flow. The spiracle opening is guarded by two lips, beyond

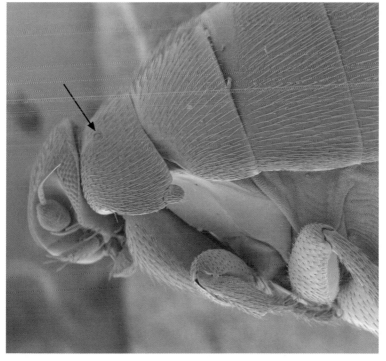

A wingless female of the genus *Epicnemis* (Phoridae) where the anterior spiracle is located on the dorsal surface.

which is a chamber, or atrium, and at the end of that, an inner valve. Beyond this valve is the respiratory system, also known as the tracheal system, which branches out through the fly's body. The upper branch provides oxygen to the thorax and off round the rest of the body, and the lower branch provides oxygen to the head. The anterior spiracle is usually up by the prothorax, the first segment of the thorax just behind two bulges called the postpronotum.

The posterior spiracle, the one positioned nearer the abdomen, has a wider opening, once again, it is protected by hairs, and two filtering lips that open and close. The rear lip in *Calliphora vicina* is fixed but the front lip is special – it is attached to the peritreme by a hinge, enabling it to seal shut or open widely. This can be opened and closed by air currents created by the changing shape of the thorax as the wings beat, causing it to act like a pump, allowing air to enter the anterior spiracle and flush oxygen through the body.

Breathing is not just a passive process brought about by the wings moving but can be more active when times require it. Muscles attached to the lips of the spiracles can alter the size of the opening thus letting more or less air in. These changes are determined by local changes, say a chemical stimulus, which results in an autonomic nervous response. One example is when a fly is working hard to take in oxygen faster than it can get rid of the waste product carbon dioxide. The subsequent build-up of carbon dioxide results in fewer nerve impulses going to the spiracle muscles, resulting in the closure muscles relaxing. This opens the spiracles fully, enabling more oxygen to enter. The opposite happens if there is a rapid, negative change in the internal water balance – in this case, the spiracle closes to prevent water loss.

This breathing system has enabled some remarkable feats in the dipteran world. For example, *Aedes euris* (Culicidae) has been recorded at 3,133 m (10,279 ft) up in the Andean mountains. Now, at these heights there is an acute lack of oxygen and secondly, the lack of air means lower air pressure for the wings to beat against, so it's trickier to actually fly. And, without being told off for stating the obvious, it's really cold! But, thanks to their remarkable breathing apparatus, they manage it. Another species, *Drosophila ananassae* can also survive at high altitudes and has been collected at an impressive 5,123 m (16,808 ft) at Badrinath, in India. But the highest recorded flier of all must be the adventurous house fly, *Musca domestica*, our most constant companion, who has once more set up home with us and made itself comfortable at Everest base camp – at 5,360 m (17,585 ft).

The thorax is a hub of activity for adult flies, home to the flight muscles and also the receptors to keep the fly level, helping it move to more favourable conditions and away from danger. For such a small body part there is so much to explore – the distribution of the hairs and bristles alone is a jigsaw that may occupy us for years. But the flies don't worry about this. Instead they carry on adapting to new challenges, new habitats and developing new morphological features, as they always have.

CHAPTER 6

The wings

If you were born without wings, do nothing to prevent them from growing.
Coco Chanel

I AM OFTEN ASKED why flies choose to fly around the living room light, and why they do so at odd angles – almost like they are flying round a box when more often than not the lightshade is circular. Really it has nothing to do with us, and instead it is to do with their flying predators, the birds, and the fly's long inbuilt ability to avoid predation. The reason they can usually avoid your misguided attempts to whack them with a newspaper is due to their fantastic vision and their ability to quickly fly out of the way. Flying is of course one of the most defining traits of a fly. I mean in English we call them flies because they 'fly', right? They are always there buzzing around – around our lights, our food and our bedrooms. From the highest mountains to a deeply buried corpse, flies have got there, and got there fast. Insects were the world's first flyers; they had mastered this feat 100 million years before the flying reptiles, the pterosaurs, had got around to joining them in the skies. The evolution of the wings coupled with their small size is thought to be the most important factor contributing to the success of all insects, but especially the flies.

The peacock fly, *Callopistromyia annulipes,* with its wings raised in the posturing position.

How one wears one's wings, however large or small, differs between the insects. Look at a dragonfly when it is sitting on a leaf and you will see that the wings are held perpendicular to the body, so the adult cannot snap them out of the way to go exploring under bark, say, or in the soil, without causing damage to these delicate structures. But the majority of the other flying insects got clever and developed a mechanism to enable them to move the wings out of the way, holding them flat along their abdomen when at rest. Some of these insects, including the flies, have a musculature attachment to a hardened plate at the base of the wing called the third axillary sclerite, and it is this that enables them to flex the wing backwards to fold across their abdomen. Back in the day, when I was a part-time lecturer, I loved teaching this subject. In my youthful enthusiasm I would flap around the lecture room as a demonstration of such mechanisms and would often see students in exams trying to copy my movements to remind them of the action.

The wings of flies, and other flying insects, are very different to that of bats and birds, being more like a hardened membranous bag that has been stretched and secured in place by internal scaffolding, rather than skin on fingers. They are incredibly lightweight and strong. How wings evolved, and from what structures, has still not been resolved, due to very limited fossil evidence to help us understand what evolutionary process occurred. It is thought wings evolved only once in insects, and that was some time around 370–330 million years ago during the Upper Devonian or Lower Carboniferous resulting in at least 10 orders of flying insects around 300 million years ago.

There are three main theories as to how wings evolved, and all enjoy various levels of support among scientists. The first is called the tergal origin hypothesis, or paranotal lobe theory, which suggests lobes on the thorax developed into wings – any features that stick out on the body would act to slow the insect down as it falls. Insect wings could have evolved from these lobes, acting as parachutes for gliding and then to fully flying. The second is the epicoxal hypothesis in which it is thought that the wings developed from a 'tracheated' gill or gill-cover, which are still to be seen on some immature aquatic insects – for example mayfly nymphs (the immature stages) have muscles to enable them to move to aid breathing and so it has been proposed that these could have gradually been modified to become wings. The third is the pleural origin hypothesis, which has many variations, but the most favoured one is the endite-exite theory, which suggests the wings developed from outer (exite) and inner (endite) articulated branches of limbs, which enlarged

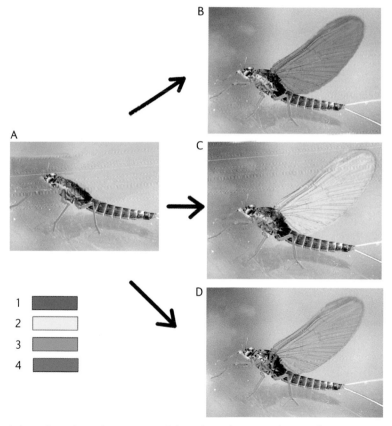

A, hypothetical wingless ancestor; B, hypothetical insect with wings from notum (paranotal-theory); C, hypothetical insect with wings from pleurum; D hypothetical insect with wings from leg exit (epicoxal-theory).

and fused together to become a wing blade. It is proposed that the leg segments closest to the body fused back into the thorax, becoming the axillary sclerites. Both the tergal and the pleural hypotheses disagree with some of the genetic evidence and developmental studies, and recent work has found that there may have been a dual origin of wings, which would unify both theories. A 2017 paper by Jakub Prokob and colleagues provides evidence for this. It argues for the formation of wings from the bumps on top of the thorax as well as from elements of a 'leg' at the side of the thorax.

Whatever theory turns out to be correct, or if any do, flies have wings but fewer than the majority of other insects. Flies have just one pair of wings, most other flying insects have two pairs, and these are located on the mesothorax. They are membranous, stiffened structures linked to the body through the axillary area – a region where sclerites are linked to the trailing edge of the wing by lobes, the size and type of which vary.

Unlike the hemimetabolous insects, who don't pupate and whose wings develop throughout their adult life, flies don't develop their wings

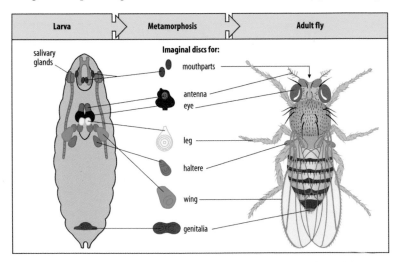

The imaginal discs in *Drosophila melanogaster* and the corresponding adult structures.

in the larval stage. Instead, inside the larva, there are pairs of cell clusters called imaginal discs. These imaginal discs are sac-like structures and are found throughout the body. For example, there is a pair for each pair of legs, a pair for the halteres and a pair for the wings (there are 19 in total all paired apart from the genitalia). These discs are nests of cells which, during pupation, divide wildly (in *Drosophila* each cell divides every 10 hours) into all of these exciting external structures. During metamorphosis these discs evert, with the legs, wings and so on, growing and expanding from the newly forming creature into the free space in the pupae. In an ever so creepy-fashion, these discs begin to fuse with their pairs to develop into one continuous epithelium, a process that starts happening around 6 hours after pupation in *Drosophila* – these discs are pieces of a living 3-dimensional jigsaw that is gradually being assembled into the adult fly.

As this development is occurring within a pupal case, the wings, although developing, are not hardened or strengthened at this stage; this happens after they emerge. Of Course there is an exception, both the Blepahriceridae and Deuterophlebiidae have their wings expanded and prepped for takeoff before they leave the pupae – these species are on a mission! Fly wings would be flimsy and floppy without their internal scaffolding, the veins. You can see these veins as lines criss-crossing the transparent membrane, where the branching of the veins allows folding, which is a crucial aid in the mechanics of flying. These veins spread and branch across the wings, providing strength and rigidity, as well as resistance to twisting. Flies do not have the incredible box-like patterns of multiple horizontal and vertical veins as seen particularly on the wings of dragonflies and lacewings, and to a lesser extent in Hymenoptera, but they do have some cross-veins, and this varies across the order. There is generally a progressive reduction in the number of veins from the lower to the higher flies, though there are always exceptions.

Take, for example, the families Cecidomyiidae (the gall midges), Phoridae (the scuttle flies) and Sphaeroceridae (the lesser dung flies). These families evolved at different times, but all the species within them have very few veins. Compare and contrast this with the families Tipulidae (the crane flies), Nemestrinidae (the tangle-vein flies) and Tachinidae (the parasitic flies). Again, these families evolved at different

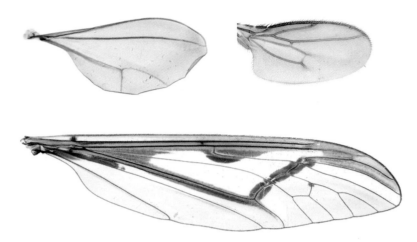

Clockwise from top left: the simple and complicated veins of cecids (gall-midges), Sphaeroceridae (lesser dung flies) and Tupulidae (crane flies).

times, but all have more complicated venation patterns. It may be that a reduction of veins is closely related to the size of the wings, and so the level of structural support that is needed.

The veins have other functions beyond the structural. The longitudinal veins also contain tracheae, nerves and haemolymph (the wings are covered with sensory hairs that have communications that rely on the network of veins to deliver them elsewhere in the body). Between these longitudinal veins there are the much-shorter linking veins called cross-veins that offer further structural support. The pattern of all these hugely important veins is also one of the best ways to identify flies down to the level of families. I'm not going to lie, the hours and eyestrain required for the dipterist to do this often begins with a lot of swearing and ends with a lot of tears.

It is not just the myriad intricate vein patterns that cause me to swear, but the names, which can vary greatly depending upon the author of the study – well, of course that's the case, it would be far too logical to settle on one universal system! The veins of insects were first named in 1886, thanks to the work of an Austrian entomologist,

Josef Redtenbacher (1856–1926). Redtenbacher went on to comment of his own work that he felt he had been 'misled' by earlier studies, and so he, and many others, carried on working on a blue-print for insect veins. Finally, in 1898, the majority of insect taxonomists settled on the Comstock-Needham naming system, named after two American entomologists John Comstock and George Needham. It settled on the following: from the top/anterior/leading edge of the wing are: the costa (C), followed by the subcosta (Sc), radius (R), media (M), cubitus (Cu) and anal (A) veins.

So why all the tears and swearing? Well, this system has been adapted over time to suit Dipterists, simplifying it with a numbering system. In their classic book *Flies of the British Isles*, two British entomologists, Charles Colyer and Cyril Hammond state that 'since we, however, are concerned here only with diptera, we shall find it less involved and therefore more convenient to adopt a simple method, much used by Dipterists, which is, in essence, based on numbering the longitudinal wings'. This system was 'further' improved upon in 1981 by Canadian Dipterist, Frank McAlpine, in the *Manual of Nearctic Diptera*, one of a series of must-have works for fly-fiddlers (and available online for free). It gives us yet another ground plan for wing veins, but one that is mostly used today.

But the naming saga does not end here either. I recently attended a course to learn as much as I could about how to identify flies in the family Empididae, the dagger flies, and their close relatives, the

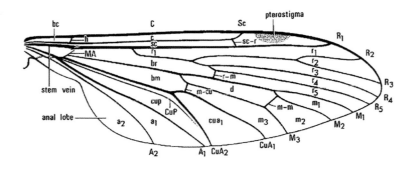

The wing stalk and blade of a hypothetical primitive diptera.

Dolichopodidae, the evocatively named long-legged flies. As I flicked through a course handout, I was met straight away with an image of a wing with yet another naming system for the veins – one adopted by Austrian entomologist Ignaz Schiner in 1864. We will need a Rosetta stone at this rate to help us decipher all the different dipteran keys to veins. This may seem irrelevant to the average reader of this book but having multiple naming systems complicates the already difficult picture of understanding which veins evolved from which, which then helps us understand the relationships between species.

Putting the naming systems aside for the moment, the majority of veins lie in the anterior part of the wing, called the remigium (which in reality is most of the wing). In the Brachycerans, the anal area of the wing (and yes, it is called that), the part closest to the body, can be very distinct, often forming a lobe, and sometimes a distinct flap called an alula. The wings change shape along distinct fold- and flexion-lines and vary in flies. The claval furrow flexes and the jugal fold, folds, and the position of these is very similar across all insects. The location of the median flexion line and the anal fold can be very variable though – even in related species of flies. These regions of flexibility and folding are very important for flight dynamics, of which more later. The costal veins, the leading-edge vein, vary across the families and some of these may be broken or merged with other veins, and these are thought to be adaptive features for flexure during flight. These breaks have sent me

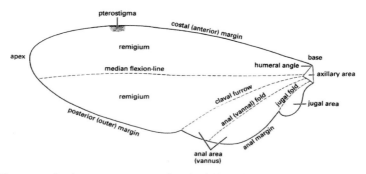

The names for the main areas, as well as the folds and margins, on a generalized fly's wing,

potty over the years in trying to determine species; they're an incredibly important taxonomic feature but one that is often so hard to see. In wasps, the costal breaks have flexible joints due to the presence of the elastic protein resilin, which enables the wing to flex and crumple on impact, but to spring back afterwards, thus mitigating against any damage. Although flies have so far not been shown to have resilin on the wings, the areas of flexibility, albeit less than the wasp, may indeed offer the same mechanical advantage.

In some flies, especially the calyptrates, including our friendly house fly, there are two further lobes or flaps called calypters (or squamae) – hence the name. These are the outer (or upper or distal) calypter that arises from the wing base behind the third axillary sclerite and the inner (or lower or proximal) calypter that arise from the very basal part of the segment, where the wings are attached. For such diminutive creatures they have amassed a lot of names for their body parts.

The calypters have long been a subject of mystery amongst Dipterists. I would often ask students during lectures about what they were for, and wait for their faces to drop, watching as they internally panicked that they had missed an important part of the lecture. I would then sneakily reveal that we didn't know either. These features are connected to the third and fourth axillary sclerites, and because these axillary sclerites are involved in the manipulation of the wing for steering, it has been suggested that calypters have an aerodynamic function, protecting the halteres from air turbulence to ensure they are still able to maintain body position. British zoologist John Pringle, proposed this back in 1948, but to this date this idea has not been investigated. Indeed, in many of the Muscomorpha, the lower calypter often forms a cup-like hood around the haltere – could this be evidence of the protective nature of this feature?

The Oestridae family – the bot flies – all have very large calypters, especially the lower calypter, which appears to stick out like an ear rather than cup the halteres. Bot flies may look slightly cumbersome, but these beasties are quite the aviators – after all, they have an awful lot of land to cover to find a suitable host for their larva to develop in (mammals do tend to roam around a lot). One experiment involved tethering two different species of botfly to a stick (this apparently happens a lot with flies). The stick that the victims – *Hypoderma tarandi* and *Cephenemyia trompe* – were attached to was fixed to a watchmakers' jewelled ball-and-socket joint.

The flies could rotate around in any direction but couldn't land. The results were quite something, but the average total flight time for both sexes and species was 20 hours, with the females of *H. tarandi* able to fly for 30 hours. The longest recorded single flight was for 12 hours straight! The researchers, Arne Nilssen (Tromsø University Museum) and John Anderson (University of California), determined that this meant they were likely capable of travelling up to 900 km (1,448 miles) in their lifetimes, i.e. greater than travelling from London to Paris and back. These two species rely on reindeer as hosts for their larvae; the former is a warble, or subdermal parasite, whose larvae develop under the skin, whilst the larvae of the latter are squirted into the nostrils of the reindeer. Both species' larvae leave their hosts to pupate, and so the adult fly may emerge some distance away from the mobile host. More importantly, they also perform some great acrobatics to ensure their larvae land in the right places on/in the mammal's body.

Another group of flies with huge calypters – and which are also parasites, or rather parasitoids, as the hosts die – are the hunchback flies, the Acroceridae, also called the spider-killing flies. These have some of the largest calypters in relation to body size; for example, the genus *Ogcodes*, where both the males and the females have enormous lower calypters. These families have a few differences – oestrid adults find hosts for their parasitic larvae while acrocerids don't, and oestrid adults don't feed while some acrocerid adults do. But morphologically they are quite similar, although the acrocerids use their modified egg-laying tube to fire off larvae like a semi-automatic rifle. Both are definitely amongst the more rotund of flies, and so perhaps the larger calypters help them fly their corpulent figures around.

Instead of the second pair of wings that most other insects have, flies have halteres. Halteres don't contribute to generating lift but instead they are essential in determining the inertial reactions experienced when flying – the movement of the fly whilst flying. William Derham, a

A male *Ogcodes* species with enormous lower calypters. The entire length of the fly is 4.2 mm.

17th century English clergyman, philosopher and scientist, was the first to determine that the halteres were sensory organs, essential for flight stability (interestingly, he was also the first person to accurately estimate the speed of sound). During flight, the halteres move at the same frequency as the wings, albeit in the opposite direction. This enables the campaniform sensilla, the stress and the strain receptors located primarily at the base of the haltere, to detect and adjust the body's rotation. These mechanosensory cells, of which there may be many (e.g. the base of the halteres of *Calliphora vicina* (Calliphoridae) have 380 mechanoreceptors, the majority being campaniform sensilla), detect the deformations on the cuticle caused by the halteres movements.

These oscillating clubs, vibrating gyroscopes, rotating dumbbells, however you'd like to describe them, are derived from the hindwings we see in the other insects. Contrary to popular belief, halteres are not unique to diptera. Most entomologists have heard of the twisted-wing insects, the Strepsiptera, an order in which the males have a pair of halteres in front of their wings (unlike behind in flies). But halteres have also evolved at least seven times in insects, with extant species including some mayflies, the males of some scale bugs, and some plant hoppers. This reduction to just one pair of wings is obviously a good idea and many of the other insects couple their wings together to basically produce one large wing – for example some insects (Hymenoptera and some Trichoptera (caddisflies)) hold their wings together by little hooks called hamuli and in some insects the jugal lobe of the forewing overlaps the hind wing, effectively forming a continuous surface (e.g. some Lepidoptera) to name but two methods.

The name haltere stems from the Greek 'hallesthai' meaning to leap. In ancient Greece, long-jumpers used to throw a pair of weights behind them as they jumped – an action that propelled them further forward. The spelling haltere was more common in British publications, with halter the more usual spelling in American publications, but for some reason the (incidentally incorrect), British spelling is the one has become globally established.

The halteres are an important component of how the fly steers. The information gleaned from the halteres and from further

The halteres of the crane flies (family Tipulidae) can be exceptionally long, replacing the hind wings seen in other non-dipteran insects.

mechanoreceptors in the wing, combined with information from the antennae, the eyes and the ocelli, all work together to give a comprehensive and incredibly fast flight response to environmental stimuli. The fly's trajectory is maintained by steering muscles, which are, in the majority, directly inserted into the base of the wing and neighbouring sclerites. This is very different to say a bird, which has muscles and joints along its leading edge. With flies this steering mechanism provides enough control to enable flies to cruise, roll, pitch left or right (yaw) as well as hover while in flight.

The steering muscles form part of the thoracic muscles, the majority of which are huge, powerful muscles that are employed together to get the fly airborne on the right trajectory. To fly, the main challenge is of course to overcome gravity – and to do this the first thing a fly must do is to flap its wings fast enough that it can overcome its weight. In winged diptera, as with all flying insects, the thorax is completely stuffed with muscles, but the arrangement of these muscles differs with orders. For example, the thoracic muscles of dragonflies and mayflies don't just power the wings, but unlike flies, they also control flight movement; these so called 'direct flight muscles' are attached directly to the base of the wings and allow these types of insect effectively to row at speed through the viscous fluid that we call air. In the dragonflies, damselflies and mayflies, these muscles contract with every nerve impulse, with a pause after each contraction so the nerves can recharge. But flies, along with nearly three quarters of the rest of the insects, have opted for a different method; instead of relying on a single nerve transmission, they operate by fibrillation or, more simply put, by quivering! These type of flight muscles are called indirect flight muscles because they are not directly attached to the wing but to the thorax. The wings beat by the muscles pulling directly on the thorax, causing a series of deformations of the cuticle – specifically the deformation of the tergum (back of the thorax). The thoracic indirect flight muscles that power the fly's wings comprise dorsolongitudinal muscles (DLM) and dorsoventral (DVM), or tergosternal, muscles.

The DLMs squeeze the thorax along the horizontal axis, which results in the thorax bulging upwards and so forcing the wings down. The DVMs squeeze the thorax vertically and so pulls the wings back up again. This antagonistic movement is not an up-and-down flapping, but a back-and-forth motion, the wings acting like little propellers around their base, before flipping back the other way. The elastic thorax stores and then releases energy during wing deceleration and subsequent acceleration – acting as a highly efficient resonance box.

The story for better flight does not stop here for flies (or the bees and beetles) for the development of indirect flight muscles enabled a further funky adaptation to happen, the development of asynchronous flight muscles. Instead of relying on a single nerve impulse to cause a contraction to start and finish, as the synchronous muscles do, one nerve impulse instead results in many contractions as the muscles now kept activated by their own increased tension that has resulted from the oscillating thorax.

Robert Hooke, the great English polymath, philosopher and architect wrote about wing-flapping in 1665 in what is, to my mind, one of the best books ever published: *Micrographia, or some physiological descriptions of minute bodies made by magnifying glasses, with observations and inquiries thereupon*. Within this mighty tome he wrote that 'the wings of all kinds of insects, are, for the most part, very beautiful objects, and afford no less pleasing an object to the mind to speculate upon, than to the eye to behold'. The man was obviously mostly referring to flies (I may be inferring bias here) and he used to study tethered flies to watch their flying motion under the microscope (it seems that we have been tethering flies for hundreds of years!). Hooke described the way flies 'flap' their wings in terms of sound, comparing it to 'the vibration of a musical string'. As his friend the MP and writer Samuel Pepys was to record on 8 August in 1666, after a chance meeting on the street: '[Hooke] is able to tell how many strokes a fly makes with her wings (those flies that hum in their flying) by the note that it answers to in musique'. Hooke guessed that the fly was 'making many hundreds, if not some thousands of vibrations in a single minute of time'. Although these observations were made more than 350 years ago, they were relatively accurate.

Olavi Sotavalta, a Finnish entomologist who moved to the USA, worked on the audio frequencies of flies and in 1953 was the first to record wing stroke frequencies of males from the families Chironomidae (non-biting midges) and Ceratopogonidae (biting midges) at a remarkable 1,000 strokes per second. Sotavalta describes the midge flight muscles as the most 'metabolically active tissues in nature'. Sotavalta states that 'a minute size is a *conditio sine qua non* for an extremely high wing-stroke frequency' – such high frequencies could not have been achieved by larger animals as there is a cost-benefit trade off associated with such rapid flight. Ceratopogonidae are some of the smallest species of flies, and this is why the Americans refer to them as 'no-see-ums'. For such small creatures to push through the air, they need to generate a lot of lift to keep them moving. Sotavalta recorded individuals from the genus *Forcipomyia* managing to attain wing beat frequencies of 1,046 wing beats per second; a sparrow in comparison flies at around 15 beats a second.

The rapid flapping of the wings of flies and similar insects have caused scientists a headache for a while. At first, engineers tried to compare fly wings to how an aircraft's wing functions. But they don't function like static aircraft wings, or even birds' wings. Instead the very high angle of attack of their wings, as they flap and rotate, causes the leading edge of the wing to make enough force to enable the flies to get and stay airborne – their wings create what is called a leading edge vortex (LEV) that creates enough turbulence that allows the fly to propel along. These LEVs form continuously as the wing is flapped, pulling the surrounding air inwards and downwards to create lift.

The wonders of wings don't even end here. In *Drosophila*, they have found that along the anterior margin where these vortices are being generated there is a line of mechanosensory and chemosensory – gustatory (taste) – hairs. These gustatory hairs are very different to the short, robust mechanoreceptors, and are instead long and thin hairs that are connected to gustatory sensilla. Although there is still much to learn about these hairs, in 2016 Hussein Raad of the University of Nice Sophia Antipolis, and co-authors wrote that these hairs responded to both bitter and sweet molecules in the atmosphere. When the flies were genetically modified to have these sensory cells silenced, the researchers found differences in both the food that they subsequently ate and their ability to explore their environment for food. They concluded that the wing-flapping process acts as a nebulizer, with the resultant fine mist of volatile molecules now being detected by their wing sensilla, helping them detect food sources without landing. Is there no end to the amount of multitasking that is undertaken by a fly's body? Flies have been described as simple creatures but is that fair? Maybe it's better to say that they utilize what they have very efficiently.

Different species of flies have both a different number and arrangement of flight muscles, and as a result, there is a range of flying styles. Listen to dipterists talk about when they are collecting flies and they will often tell you how specific species behave in the air. For example house flies' whizz and whirl around, while bee flies hover and dart. The majority of flies though, flap their wings in the horizontal plane while the body droops down. But, the aptly named hover flies (Syrphidae), keep their bodies in a horizontal plane, instead tilting the wings. Next time you are in a garden, check them out, for they are able to hold their body absolutely straight, which is a remarkable feat.

This type of hovering creates aerodynamic instability, which should result in out-of-control flies. But both their mean centre of mass and the mean centre of pressure counteract any increased instability. These flies (and incidentally, the dragonflies) are able in this way to become the masters of hovering. Many males hover to guard their territories, or form mating leks, both of which are energy-expensive activities. Some species of fly, the smaller syrphids for example, overcome this by warming up their flight muscles. They prepare for flight by activating their wing muscles without allowing the wings to beat, thus warming up their muscles first – a more energy efficient system.

All adult hover flies are very dependent on nectar and honeydew for carbohydrates, e.g. glucose, fructose, sucrose, to provide enough energy

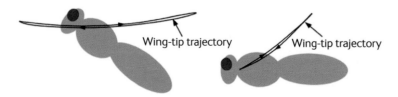

The two types of hovering in insects – normal and inclined-stroke-plane hovering.

for flight and other metabolically important functions. But intensive agriculture has led to a reduction in these floral reserves. This is a great loss to us as well as the fly. For example, the colourful *Episyrphus balteatus* are very important pollinators, but also have highly predacious larvae, preying on many agriculturally important pests such as aphids. Not only are we destroying the pests, but we are also destroying the pollinators and significant biological control agents. Understanding the energetic requirements of these important species is an understudied but essential topic.

A further adaptation that makes flying very efficient (and is found in dragonflies as well), is the highly elastic protein resilin. Resilin can be found tucked between the direct flight muscles and the wing and gives an extra rubbery flexibility to the wing joint. It's outstanding – firstly because of how efficient it is – very little energy is lost through heat while it's being worked; and secondly, how resilient it is (the clue is in the name). This protein is incredibly durable, and it has to be considering how much use it gets. In 2011, American entomologist William Wiesenborn discovered that resilin is also found in halteres, enabling them to retain their shape despite all the movement.

Flies don't just fly, there is a lot of walking and running to be had. The legs are the chosen method for locomotion for these movements, but the fly still needs the halteres. If you were to cut the halteres off a fly (and in no way would I encourage you to) you would cut off its ability not only to fly properly, but with some species, you would also affect their ability to walk. Kathryn Daltorio and Jessica Fox, of Case Western Reserve University, USA, conducted experiments in which they removed the halteres from *Sarcophaga bullata* (Sarcophagidae) and found that these poor beasts had massive problems in coping with walls. On both horizontal or vertical surfaces, the speed decreased with the haltere-less flies, even more so on the vertical surfaces – the sides of cups in this case. Many individuals would not even attempt to scale the cups, whilst others did and then fell off.

It is only those flies that oscillate their halteres while walking that are affected. Sarcophagids are amongst those flies, as are other species in the calyptrate families Calliphoridae, Muscidae, Tachinidae and

Anthomyiidae, and acalyptrate family Micropezidae. Those families that don't waggle their halteres around are not as adversely affected; for instance, *Drosophila* don't waggle, and manage to walk fine without their halteres. Daltorio and Fox hope that this sort of study, i.e. showing the multifunctionality of halteres, will help with the development of biosensors, enabling the development of sensors to help humans with sensory disorders, or machines to land better.

Thanks to all of these structures and systems working together, flies have been able to travel great distances for both food and sex. Flies such as the wonderfully metallic bluebottles are often the first at the scene when fresh food arrives. Some flies are able to migrate vast distances – even across Europe or the United States – as with the hover fly *Episyrphus balteatus,* while other flies are among the fastest of the airborne insects. According to the Smithsonian website the fastest insect is a dragonfly, at 56 kmph (35 mph). But some flies are thought to be faster. How fast? Well, there is a still lingering myth that one fly can break the sound barrier. I like to think the persistence of this fable could be something to do with the fact that the supposed record-breaker was a beautifully hirsute and rotund deer botfly, *Cephenemyia pratti,* an alluring creature that never fails to make folks fawn at their fluffiness. Or maybe it's just because it has such a fantastical speed that we want to believe it is true.

The source of these rumours was one Charles Townsend, the American Tachinid specialist that I talked about in a previous chapter who, in 1927, reported that according to measurements he had taken in the field this particular fly had achieved the incredible speed of 400 yards per second (that's 818 mph or 1,316 kmh – or around 50% faster than a jet airliner). Yes, apparently, he was able to calculate the speed from his estimates of an individual rushing past him. Townsend's findings were only finally refuted in 1938 by the spoilsport Nobel Laureate for chemistry, Irving Langmuir, who just could not accept this figure, and quelle surprise, disproved it. Townsend was already known for courting controversy, even for his taxonomic work and was (and still is) criticized by fellow dipterists. As Canadian Dipterist James O'Hara, wrote in 2013: 'Townsend's methods and productivity are worth more than a cursory mention because this author has, in some ways, done more to retard tachinid taxonomy than advance it.' Townsend had a habit of offering his opinion regardless of accuracy. In his *Manual of Myiology* (1934) he regularly digressed from the subject at hand into discourse about the origin of the moon, human development and gravity. The speeds achieved by this species have now been corrected to a far more reasonable – but still pretty fast – 25 mph (40 kmph).

That said, a male horse fly, *Hybomitra hinei wrighti*, has apparently been clocked at 145 kmph (90 mph) – while chasing a female of course. Jason Byrd tells us this in his 1994 book *Insect Records*, while telling the story of a bizarre experiment relayed to him by Jerry Butler, a medical and veterinary entomologist at the University of Florida, USA, who has worked with many economically important insects.

Butler, and his University of Florida colleague Richard Wilkerson, had undertaken this experiment 10 years previously. In arguably one of the most amusing insect trials to date, Wilkerson and Butler modified a Crossman pump air gun to shoot black beads, intended to look like female horse flies, above the heads of the males, and then filmed the way the males flew in response. As well as being able to discern how quickly they were moving, they found that the flies always performed a specific manoeuvre to pursue the black beads zooming over them. Because the move was similar to one developed by German First World War fighter pilot Max Immelmann, they named it after him. Immelmann would pull his plane up into a half loop, and then half roll it, increasing both height and then rapidly reversing direction. The same Immelmann turns were being performed by the excited males, and at an incredible speed.

So, while wings are primarily for flight, they are also major components in a male's toolbox of wooing devices. Some lovely work has come out of understanding how courting male *Drosophila* 'sing' to females by vibrating their wings. Each species has a different repertoire, from loud purrs to soft drones, with some males able to alter their songs depending on the reaction of the females. If receptive, the females allow their genitalia to be licked, and copulation occurs. He sings to slow her down and encourage her to be more receptive to him. This is not always successful – often she rejects him and produces a rejection sound – she tells him to 'buzz' off and may aim a kick at him.

A very battered *Nowickia ferox* (Tachinidae).

Wing-waving is another common trait amongst flirty males, and one of the larger and more splendid of the UK species that do this are in the Dolichopodidae family, specifically *Poecilobothrus nobilitatus* – the semaphore fly. The male, as the name suggests, signals by waving his white-tipped wings perpendicular to his body while on the ground to attract the attention of nearby females. Many others in the family perform their courtship routines in the air and are able to perform very dynamic flight manoeuvres as a result. Wing patterns vary widely across the family, too, hinting at how important it is for this family to have great manoeuvrability and speed for hunting and courtship – but that it is also important to look good.

Courtship, along with hunting, guarding territory and dispersal into new habitats, all takes a toll on the wings. At the end of the breeding season, a shabby wing is a common sight, with many individuals sporting torn or tattered edges. Researchers have found that foraging bumblebees collide with plants on average of 60 times a minute – once a second – and this can lead to a loss of 5–10% of wing in just one day! I am not implying that the humble bumbles are the clumsiest but that does seem quite extreme. Flies spend a lot of time on the wing and foraging amongst the flowers and as such experience similar wing damage. Flies have to balance wing manoeuvrability with support but at the end of the mating season there are many casualties.

Hover flies, those extreme flies, have very rigid lower veins that join to form a toughened lower edge, often called a false edge, which may limit tearing that is seen in so many of the other large fliers. This, along with the presence of a vena spuria – a false vein – are diagnostic features of this family, indeed, the latter are only found in this family. Because the vena spuria is an 'apomorphy', i.e. not seen in any other families of flies, we are unsure as to its origin. There is another family that has a diagonal vein – actually a series of veins – that combine to form a strong single vein. These are the tangle-vein flies (Nemestrinidae), and, as the name suggests, have some of the most unusual veins of the flies – an arrangement that, in many species of this family are referred to as 'reticulate' though not all nemestrinids have reticulated veins but they do all share in common this diagonal 'vein', so this is a diagnostic characteristic for the family. The vein comprises elements of R, M and CU. The vena spuria and the diagonal 'vein' could be what are called 'analogous' – structures that have different

The highly reticulate veins of *Nemestrinus signatus* and the strong diagonal 'vein'.

evolutionary ancestors but evolved separately to do the same thing; both syrphids and nemestrinids are great hoverers. The position of the veins affect the aerodynamic properties of the wing and it may be that this vein is beneficial to this type of flight.

Whatever the origin of these veins, or even all the other primary and secondary veins, may be, their position on the wing is very helpful in understanding the evolutionary relationships between species. Incredibly there are not only differences in the patterns across species, but in the pattern between the left and the right wing of the same species, as Jordan Hoffmann and co-authors found in 2018 – many combinations can end up having the same effect even though they look different. The vein thickness and arrangement, and the number and placement of cross veins, can all affect how well a wing operates, as well as of course shape, size and muscle attachments – a multitude of factors that, combined, create the fantastic flying machines that we see around us.

Wings are not always flat, in fact, many are corrugated, and we can see this feature easily in the blow flies. This slight folding along the surface does not appear to have any obvious aeronautical advantages, as far as we know. But it has obvious structural advantages in making the wings more rigid without any weighty additions. The corrugations look pretty too, making stunning reflections as they catch the sunlight.

Patterns, shapes and colours such as these are crucial to the mating behaviours and very survival of many species of fly. In many flies, the wings may be patterned with spots, stripes, and of course, those intricate veins. Many species will use these as amazing disguises to mimic their prey and catch it unawares, while others will use this mimicry for protection. For instance, there are many species that take on the appearance of better-protected species, many of the hover flies resemble wasps and hornets. And there are even one or two that have evolved to look like one of their main predators – spiders.

One such species is the snowberry fly – like a sheep in wolves' clothing, as authors Monica Mather and Bernard Roitberg, at Simon Fraser University in Burnaby, British Columbia, put it in their 1987 paper of the same name. The authors describe in detail the mimicry behaviour of this species, *Rhagoletis zephyria* (a Tephritid), which occurs throughout North America, living on snowberry bushes and generally trying to avoid being eaten. Its main predator is the zebra spider, *Salticus scenicus*, a very territorial jumping spider (Salticidae), which hunts by stalking its prey. Now, the snowberry fly's wings are patterned in such a way that it is able to mimic the spider's legs when

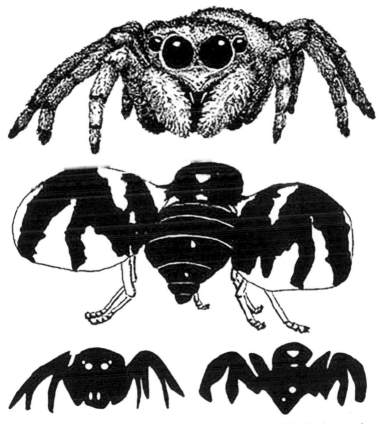

The front view of the zebra spider, *Salticus scenicus* (top) and the back view of the snow berry fly, *Rhagoletis zephyria* (middle). The bottom diagram shows the dark parts of the animals (left, spider; right, fly).

in a raised, defensive position, What's more, the plucky little fly can also perform a side-to-side dance that is almost identical to the movement of the spider when it's in its defensive position.

The clever *Rhagoletis zephyria* also uses its markings in the pursuit of romance. Many species of tephritids use specialized dance moves to attract the opposite sex, and a vocabulary has even been developed to describe the many different types of these wing movements used during courtship that include arching, where the wing tips nearly touch the substrate, and lofting, where the wings are held parallel to each other above the back of the fly.

The Tephritidae belong in the superfamily Tephritoidea, a group of flies that we informally call the picture-winged flies. All seven families in this superfamily contain members with wings with obvious markings and in one of these families, Ulidiidae, there is one of the most spectacular species, both because of its wings and what it does with them. The peacock fly, *Callopistromyia annulipes*, is a stunning species. Their bodies and wings are gorgeously mottled, which are used by the males to attract a mate by raising them over their bodies, into the lofting position, and then tapping them together – a position known as posturing. Incidentally, it usually is the males that are most patterned as these are the sex that does most of the flirting.

What humans see in these wing patterns is only part of the picture. There is a whole world of colour available in ultraviolet (UV), a part of the light spectrum that we cannot see, but flies can. Thin wings reflect very vivid UV colour patterns, wherein reflections of the lower and the upper surfaces alter the light by either enhancing or reducing it altogether – a phenomenon called thin-film interference. An example that we can see of this in everyday life is on soap bubbles. On flies, these colour patterns vary with species depending on the thickness, pigmentation, venation and distribution of hairs on the wings. By studying these patterns, we are beginning to be able to identify 'cryptic' species – those capable of disguise.

Finally, I can't write a whole chapter about wings without a special mention of the flies that don't fly. Some flies don't fly, and these are known as apterous species. This state may exist for the whole of an adult's

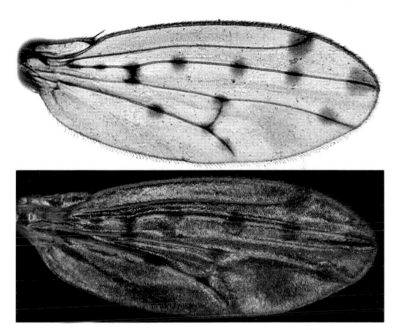

The right wing of holotype of *Drosophila guttifera* (Drosophilidae).

life or just part of it. Some have evolved that way to adapt to specific environments, while others lose their wings in various life processes. One extreme example of a fly losing its wings during its lifetime is in the genus *Ascodipteron*, from the Streblidae family, a group of ectoparasitic bat flies. The female of this species rips her wings off after copulation. I am very lucky to work on these flies, but I have yet to see a mated female. And the reason is this – the fully-winged female hides herself in her bat host after copulation. Utterly amazingly, once inside, she not only rips her wings off, but her legs as well, and basically undergoes a second metamorphosis inside that reduces her fly-like adult form to one that is just a reproductive sac. Theodor Adensamer, an Austria-German entomologist, first described this genus, found exclusively in the Old World (Africa, Asia and Europe) in 1896. Adensamer's observations understandably focused fairly intently on these incredibly unusual

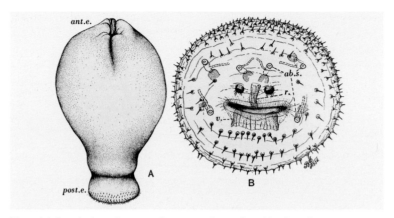

The adult female *Ascodipteron africanum* – her reduced body and missing (invaginated) head (A), and the exterior posterior end (B) (not a smiley face).

females. He described the first species, *Ascodipteron phyllorhinae,* from a single specimen that was found in a bat from Indonesia, writing that the specimen 'is bilaterally symmetrical, bottle-shaped and shows no segmentation'. Most of the female was buried under the skin of the bat, leaving nothing but a 'button-like protrusion' poking out. He concluded that this was the posterior end. In fact, the mouthparts and thorax also collapse into the abdomen and the posterior end.

This whole Streblidae family is weird when it comes to morphology: – you can also find among them stenopterous (narrow) wings, brachypterous (short or reduced) wings and macropterous (long or large) wings. This sort of wing modification is commonly seen in flies that are ectoparasites of bats and birds as they simply have no need for them. But they aren't the only ones. The tiny males (1.5 mm or 0.2 in) of a flightless midge of the genus *Pontomyia* in the Chironomidae family, have short, stenopterous wings, which are no longer any use for flying – instead, they use these for rowing, for these are now a marine species. The male's wings have evolved into fleshy oars, which they use to row around on the surface of the ocean, and the males need all the help they can get to paddle around and find the females in time. They also have very long, thin legs for stability, as well as a very long antennae and well-developed eyes for this purpose – the females are very hard-to-spot, at just 1 mm long! The females don't even emerge from the water. Instead, she pokes her tiny genitalia through the ocean's surface, as she has no need to fully emerge as she is both wingless and legless. And so, the poor males have to scoot around trying to locate the female's tiny genitalia bobbing on the surface like a small buoy. The species in this genus are remarkable for several things. The adult males have an incredibly short life-span – one of the shortest of all insects at just one to three hours. It's even shorter for the female – she has but 30 minutes to complete her life and becoming a mother.

The adult males of the genus *Pontomyia* (Chironomidae) with their long fore and hind legs and their paddle-like wings, photographed in Pulau Tekukor, Singapore.

Another member of the Chironomidae family, there are so many weird species, is the Antarctic midge *Belgica antarctica*. This species has no wings either and this has led to the very unusual sight of it forming not aerial swarms, but terrestrial swarms on the ground. Reaching a whopping 2 to 6 mm (0.079 to 0.24 in) in length it is the largest purely terrestrial animal of this zone. Yet incidentally, it has the smallest known insect genome known to science to date, at 99 million base pairs of nucleotides (a house fly has 691 million base pairs). And then at the other extreme in this family, there are chironomids with massive wings. And there are some right beauties, such as the extremely unusual species that live in caves, e.g. *Troglocladius hajdi*, found at 980 m (3,215 ft) below the surface in the Lukina Jama-Trojama cave system in Croatia, which I discussed in Chapter 2. It retains all the features typical of a cave insect – pale, translucent bodies and very limited vision – but with the addition of large wings. Most true troglobionts lose their wings, preferring to walk around, but this species has very large wings and it is incredibly hirsute; these hairs are thought to be sensory so preventing the flies from smashing into walls whilst it is flying.

Among my favourite wings are the big beauties of *Exsul singularis* from New Zealand – also known as the bat-winged cannibal fly. This name is a definite misnomer. They should probably just be called bat-winged flies, because there is no evidence to suggest they have cannibalistic tendencies, but I have to admit that does not sound as dramatic. There are photos of them predating on other animals – one I have seen shows one eating a butterfly, so predators yes, cannibals probably not. This species is found in the regions of Westland and Fiordland, and the males of this species are magnificent with their enormous and darkly coloured wings. The females are much more normal in appearance, with much smaller wings. Why the males have such large wings is not fully understood. Maybe they are mimicking their prey, such as some of the larger winged butterflies, or maybe once more it boils down to sexual attraction.

The diversity of species of diptera has in many ways been advanced by their ability to spread across the planet, to find new niches and exploit

A male bat-winged cannibal fly, *Exsul singularis*, from New Zealand and his amazing pantaloon wings.

new habitats. And their ability to fly has been the major contributing factor to this. With many questions still unanswered about the evolution and properties of wings, we will be kept busy with research for many more years, all of which means nothing to the fly of course.

CHAPTER 7

Legs

Feet, what do I need you for when I have wings to fly?

Frida Kahlo

WINGS ARE GREAT for most of the time. But we all need our feet firmly planted on the ground some of the time. And flies' feet don't just grab onto the ground, they grab onto all sorts of things – including their victims, their lovers and themselves. They use their six legs not just to move but to clean, to feed and to attract the opposite sex. When that (human) male waves at you from across the room to try and get your attention, think about the little fly sitting on a lily pad waving his legs hoping that he too will catch some lady's eye.

Adult flies have three pairs of legs that may not be the sturdiest, or even the most dextrous in comparison to other insects, but they have enabled flies to swim to the bottom of lakes, crawl through the ground to the corpses buried below, and wrestle with animals at least twice their size. The rather wonderful Renaissance man Robert Hooke, whose interest in flies has already been mentioned, referred to their feet as 'the most admirable and curious contrivance' with 'handsomely shaped' talons (claws).

Having three pairs of legs is one of the defining characteristics of an insect – although not all six-legged creatures are insects. Collembola (springtails), Diplura (two-tail) and Protura (first tail) all have three pairs of legs, but also have internal mouthparts (the entognathous

Campsicnemus magius and his highly decorated front legs.

species that I have mentioned before) and as such aren't considered true insects, but join them in the subphylum Hexapoda.

There is a joke among entomologists that having six legs is not compulsory for some flies i.e. the crane flies, which are notorious for losing theirs. Crane fly legs are often very brittle and are referred to amusingly as deciduous, as they are often lost from the body. This is not just because they are the feeblest of flies, but rather that they have decided to employ a defence mechanism called autotomy, a process whereby they can shed a leg if grabbed by a predator – a very handy technique for these long-legged creatures trying to avoid predation from both winged and webbed predators (but a nightmare for anyone studying or trying to maintain a collection of them). I often see adults flying around with less than a full complement, and it's not uncommon to see individuals with five, four or even three legs. These can't grow back though, unlike other autotomous species such as lizards and crayfish whose legs do. (Incidentally, in the 2018 *Guinness World Records* a giant crane fly collected by Chinese entomologist Zhao Li in 2017 at Mount Qingcheng, China, was verified as the world's biggest crane fly, *Holorusia mikado*, that spans 25.8 cm (11¼ in) between its foreleg and hindleg.)

Each pair of legs is on a separate segment of the thorax, and are named the forelegs at the front, midlegs in the middle and, hindlegs at the back. More technically, these are the pro-, meso- and metathoracic legs. What they look like varies dramatically depending on function, but they

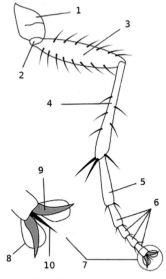

A simplified leg of a fly: 1: coxa; 2: trochanter; 3: femur; 4: tibia; 5: basitarsus or first tarsomere; 6: second, third, fourth, and fifth tarsomeres; 7: acropod; 8: pulvillus; 9: claw; 10: empodium.

all have the same basic leg segments, from body to foot: the coxa, trochanter, femur, tibia and tarsus. And on the tarsus, the foot, there are individual tarsal segments called tarsomeres. And of course, bristles can be found everywhere, sometimes a few, sometimes a complete forest of them, but I will save these little taxonomic nuggets until later in the chapter.

The coxae, also known as the base or basal segment of the leg, are the hips of the fly and these are the parts that articulate with the thorax. The coxae are generally quite short but there are exceptions, with some of the nematocerous families — mosquitoes, crane flies, gnats for instance – having incredibly swollen or

The massive elongated white coxae of *Exechia contaminata* (Mycetophilidae).

elongated coxa, including species of Mycetophilidae, the true fungus gnats. As well as possessing the massive muscle-packed thoraxes, they have very obvious coxae, some of the largest in relation to body size of all the flies. I do like these flies, as they have one of my favourite body shapes – smooth, humped thoraxes, often with their heads fitted in snuggly beneath their 'shoulder pads' and their long legs – why they have this shape is not known. The coxae are the parts of the legs that join the body and have to be flexible to allow for movement both whilst walking and flying. During flight, the legs are held differently across the order, with modifications in the coxites (the plates that make up the coxa) allowing for greater streamlining in most of the Brachycerans.

Midges, gnats and the rest of the nematocerous flies, including these mycetophilids (as well as the robberflies and their relatives), fly with their legs splayed, whilst the higher flies fly keeping their first and second pair of legs tucked into the body to keep the fly streamlined, and use the hind legs to help steer.

The legs of insects, unlike their relatives the Crustacea and Myriapoda, have gradually evolved to become closer together, and are located more underneath than at the side of the body. Adult flies, like all arthropods, have jointed legs, and locomotion in these (as in all jointed animals) involves moving all the legs in a coordinated fashion (in humans this often fails in toddlers, teenagers and the seriously inebriated). The gait of most flies, for most of the time, is known as the tripod gait. Here, three legs are in contact at any one time with the substrate, and for each pair of legs, one is up and the other is down in an alternating pattern. The legs in the air swing forward, whilst those on the ground both pull forward and provide a very stable base of support. This coordination is governed by a system of neural networks called central pattern generators, which are able to control and generate rhythmic outputs, such that all the legs are told what to do at the same time.

Meanwhile, the proprioceptors detect the position and load of the legs, and, working in tandem with the information received from the eyes, antennae, and the rest of the sensory apparatus that has already been covered in this book, the fly is able to walk steadily and respond rapidly to any change that comes over the horizon – just like the seafarer who learns to keep in three points of contact at any one time on their boat. There has been a lot of work studying the legs of some of the interesting jumping species of insect, and the internal arrangement of muscles, but oddly, even with *Drosophila,* there has been very little research into dipteran leg muscles since Albert Miller's work on leg development in 1950. This was finally rectified in 2004, when Cedric Soler and colleagues at Wroclaw University published a comprehensive study showing muscle development, together with their attachment around the internal tendons – a previously unknown feature, and one that resembles our own muscles. They found that the muscles and the tendons grow alongside each other to produce an arrangement of multifibre muscles – where many muscle fibres form a bundle that can be stimulated for movement, some of which can be very rapid. The internal leg muscles are attached to the tendons on the inside of the leg's cuticle upon which they act to cause movement. When you are munching on a crab's claw and you pull it out of the main part of the leg, you are holding what is technically called a chela. You will notice that it has a long, plastic-like structure – this is the tendon, the structure that connects the muscles to the body and in this case the claw. The Wroclaw team's work was made possible by combining two genetically modified *Drosophila melanogaster* strains (the MHC-tauGFP strain was combined with the 1151-Gal4 driver, and then these flies were crossed with the UAS-dsRED line – yes, this is the language of science), producing flies whose muscles and tendons were now visible. In other words, they

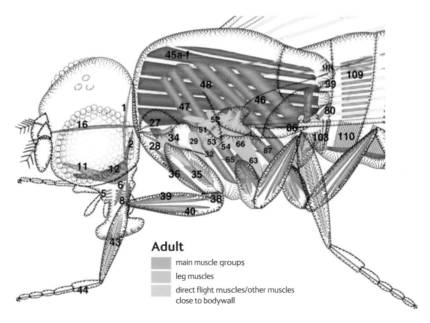

The many muscles associated with walking, swimming, running etc. in particular the internal leg muscles: 35, depressor of trochanter; 36, lateral levator of trochanter; 38, reductor of femur.

threw antibodies at these ever so slightly genetically modified types of (dead) *D. melanogaster*, which highlighted the now stained muscles and tendons under a confocal microscope. The images produced by experiments like this are a long way away from the images that Miller was able to produce, and the team was able for the first time to show the tendons running through the legs, the muscles and where they were attached to the apodemes – a ridge-like ingrowth of the cuticle. Muscles are found throughout the leg apart from the tarsal segments, where there is just one long tendon.

The movement of the leg is brought about by both the muscles in the thorax, which control bases of the legs, and the contraction/relaxation of the internal leg muscles, which either expands or contracts the leg. The muscles in the thorax involved with the movement of the leg are attached in the trochanter and coxa, and instigate the up-down movements as well as some of the rotation.

Legs are not just for walking or steering. A fabulously gruesome modification has been the development of 'raptorial' legs, a term arising from the way these species grasp prey in a vice-like grip between their femur and tibia, much like a raptor would hold its victim between its claws. A good example is the Ceratopogonidae midges. These species are not only the sanguivorous feeders that people curse, many species are predators, hunting other insects. In many predacious midge species, the female is the main protagonist (many adult males are very short lived and don't need much if any food – you will hear why shortly). She grabs on to her prey with her raptorial forelegs and then pierces them with her proboscis, releasing proteolytic saliva to dissolve their insides which she subsequently sucks out. Amusingly, she does this to the male of her own species, too, while in the process of mating!

This macabre mating ritual was first recorded back in 1839, when Rasmus Carl Staeger, a Danish dipterist, suggested this could be what was happening in *Mallochohelea nitida* midges (then called *Ceratopogon nitidus*). But it wasn't until 1914, when Maurice Goetghebuer, a Belgian Dipterist, noticed torn-off male genitalia still attached to the females, that we were able to confirm Staeger's observations. This is not such unusual behaviour for a large number of species, as there are many

A female *Mallochohelea* sp. with her raptorial front legs for grabbing her prey, which may or may not also be her lover.

flies that supplement their diet by feasting on their sexual partners. The suicidal males form swarms, into which the female enters and grabs a male to feed upon, during which he has a finite timeframe in which to successfully copulate with her. There are many records of this behaviour in the subfamily Ceratopogoninae – specifically in the tribes Heteromyiini, Sphaeromiini, and Palpomyiini. Males in these tribes are very small in comparison to females and their antennae are greatly reduced, presumably as they no longer need to find the female – she is the hunter. The other tribes in this subfamily have normal-sized males and fluffy antennae, and there are very few records of males being eaten.

The observations of sexual suicide are too frequent and the body modifications too extreme in this family for this behaviour not to be considered a systematic adaptation rather than just the odd female being confused during copulation.

Another raptorial predator, the bodybuilder of the fly world, is in the shore-fly family Ephydridae, in the genus *Ochthera*. Their forelegs are incredibly swollen, and look like bulging quadriceps, and these bad girls and boys prey on many species of mosquitoes and midges, during all but the egg stage of the prey's life.

Interestingly, as a result, a few, such as *O. chalybescens,* have been investigated as potentially useful biological controls. Noboru Minakawa and fellow researchers at Nagasaki University published a paper in 2007 on the feeding habits of this shore fly which included the mosquito *Anopheles gambiae*, the primary malarial vector in sub-Saharan African, and found that one adult shore fly could on average eat between 10 and 19 mosquito larvae over 24 hours. That's some appetite and one that could be useful to us.

Ochthera mantis was the first species to be described in this genus and in terms of looks it's the most conspicuous one, but all are commonly called mantis flies (not to be confused with mantidflies, the common

The aptly named mantis fly, *Ochthera,* with its bulging raptorial front legs.

name for Mantispidae – a neuropteran family). *Ochthera mantis* has particularly amazing forelegs, where the femur is strongly swollen, and the leg is positioned in the same way as a mantis – grasping its prey between the femur and the tibia. As American dipterist Frank Cole, described the adults' behaviour – they 'crush small-bodied flies as well as snatch mosquito larvae from the surface of the water'. But why do they need such formidable forearms for catching such weakly sclerotized prey? Well this genus doesn't just feed on mozzies and midges but other flies too, and some bugs. They also use those protibial spines (that incredibly evil-looking spine on their first tibial segment) for digging out the larvae of Chironomids. They also wave their great pins in the air to both threaten rivals and display to their own females. To help it show off, the mantis fly has bright, shiny UV-reflective surfaces on its legs, a feature also found on its face. But this may not be for flirting per se, rather one of simple recognition. In 1975, American entomologist Karl Simpson wrote that *O. mantis* and *O. tuberculata* looked very similar, but postured differently, so these displays may be more for the females simply to identify males from their own species.

Some of the most well-known grasping hunters belong to the closely related families Empididae, the dance flies, and Dolichopodidae, the long-legged flies. The Empididae consist of more than 3,000 species found around the world and are a lovely group of flies where the hunting is done by the males, and this has led to some of the most astounding mating rituals of all the flies. In the subfamily Empidinae, there are many species where the male hunts smaller insects, including many species of other flies. He does this not for himself but as nuptial gifts for the female, to persuade her that he would make a suitable mate – the better the food offering the more willing she is to copulate with. Sometimes she will grip these in her raptorial forelegs, feasting away while she and the male are copulating. Adding to the romantic scene, they are often seen dangling from a leaf, with the male holding on to her with his forelegs, supporting both the female and her feast, while she apparently pays no attention to him at all! Many of these species wrap their gifts in silk balloons, which is thought not only to further display his fitness for breeding, but also slow her feeding down to give him more time for a successful copulation – she is a quick eater and will hightail out of there even if he hasn't finished. Sneakily, some males in the family cheat. *Empis opaca* flies have been observed giving the females inedible fluffy willow seeds instead of an edible nuptial gift, but it has been noticed that their success rate is not that brilliant – cheap offers of food are seen to be just that by the majority of females.

Wrapping gifts occurs in many genera, but within the tribe Hilarini, a group commonly referred to as the balloon flies, these courtship practices are even more varied, and sometimes very sneaky. The males have secondary sexual characteristics – not involved in the direct transfer of sperm – on their swollen feet, the basitarsus, in the form of silk glands. These are similar to the spinnerets of spiders, but instead of building silken webs, these romantic males spin silken balloons as

A mating pair of *Empis aerobatica* from North America – she is preoccupied with feeding, he is preoccupied with holding on!

gifts to the females. What is quite fascinating is that the balloons have no nutritional value. This was first discovered by a pair of researchers at Macquarie University, Sydney, Australia, in 2003. James Young, a silk specialist and David Merritt, an evolutionary entomologist noticed that these silk glands, in an undescribed species they temporarily labelled Hilarempis20, were composed of clusters of diatoms (single-celled algae) crudely bound by silk thread. Nuptial gifts appeared to have gone from those that are essential for feeding to solely ones of proving prowess. It's thought that originally the silk would have been used to subdue prey, but that this evolved into the wrapping of prey balloons, which led finally to the creation of these empty or inedible balloons.

These swollen, silk-producing units are known as 'simple Class III dermal glands' and are generally only found within *Hilara* species. But in Australia another genus *Hilarempis* also sports them. Young and Merritt found that these were made up of approximately 10 pairs of secretory spines from which the silk was produced, housed in a groove on the basitarsus, which were connected to internal glands via canals. They concluded that these glands probably developed from chemical sensilla, which are common on the legs of flies, and help them 'taste the air'. They did not have to have the characteristic swollen appearance. Young and Merritt also found silk glands in males whose basitarsi were not swollen but they could still perform the same task.

Either way, these males bearing gifts, if lucky, still face a logistical nightmare when copulating. The male, sometimes whilst in mid-flight, has to support both the female and the gift, as during some of the time she is feeding, she is not actively flying. That is an incredible load that he is supporting. Next time you whinge about the demands of your partner, think about what these males have to go through – these newly married husbands have to carry their bride and presents across the threshold simultaneously.

Whether or not the male possesses these silk-producing feet does not just vary across species, but can often also vary within species. In 2010, a new and very weird Empididae species was found in Japan, by Christophe Daugeron, a Dipterist at the Muséum National d'Histoire Naturelle in France, and his team. Daugeron named it *Empis*

(Coptophlebia) jaschhoforum. This species is very odd because it has so many variable forms. Some males have no expanded tarsus at all (and they picked one of these examples for the 'type' species, i.e. the specimen that the description is based on), while some have them on the left leg, some on the right and some on both!

The reason for this, the authors hypothesized, comes down to one of two things. Firstly, it might be female choice, but then why some would choose an unadorned male is unclear – maybe they are better at catching prey as they are less hindered? Or it might be that reduced encumbrance is an advantage for males in fighting over females? Or maybe it is a combination of genetic and environmental factors – a poor larval diet or odd genes? Only research can tell.

The midlegs on some species are also enlarged. The greatly entertaining Hawaiian dipterist, Neal Evenhuis, at the Bishop Museum in Hawaii, has worked extensively on the dolichopodid genus *Campsicnemus* (among many others) in Hawaii and nearby regions. Many of the genus have developed particularly enlarged midlegs, but why the midleg has been enlarged rather than the more usual foreleg, we do not know. After

The expanded basitarsus of male *Empis (Coptophlebia) jaschhoforum.*

The female *Empis decora* – a big lady with hairy legs and silky hairs.

identifying one particularly beefed-up new species from Tahiti, he named it, of course, *Campsicnemus popeye,* a nod to the famously muscly cartoon character. What is extraordinary is that more than 200 species have been described in this genus in Hawaii and French Polynesia alone – over half of the known global species of this genus. You rarely get two or more of these species living in the same environment, but in the Pacific, Evenhuis found up to eleven in one sweep. He surmises that they must vary enough in their sexual signalling so as to not confuse females from another species as suitable mates.

It's not just the male empidids that have sexual characteristics on their legs, the females have pennate, or feather-like, hairs on their back limbs, as seen with *Empis (Coptophlebia) jaschhoforum.* And in other species in the genus, these fringes can occur on other legs – for instance, the midlegs in *Empis decora* and *Empis pennipes* – two absolute knockout species found throughout Europe including the UK.

Another rather famous female in the fly world, endowed with a set of flirting features, is the dance fly *Rhamphomyia longicauda*, again in the Empididae family. She not only has very hairy legs, but she has an abdominal sac that she can turn inside out and inflate. She then wraps her hairy legs around the sac, which males find irresistible, skills thought to emphasize her ability to produce many offspring! It's these kinds of manoeuvres that earn the species the family name of dance flies.

So why do the females have these ornaments? Most females do not have any sexual ornaments – they take up a lot of resources that could be better utilized for reproductive purposes. But in polyandrous species – where the females have multiple mates and store sperm – some females have decided to advertise their wares to get the better males. Why males want to mate with these promiscuous females may be because often the sperm from her last copulation is what she favours to fertilize her eggs. Studies have found that species have fluffy legs as honest indicators of how fecund she was, and that the males were more likely to mate with females with hairier legs. This was important for females with small abdomens, because wrapping their legs around them creates a bigger silhouette for the male to see. For species like *Rhamphomyia longicauda,* in which both big bums and hairy legs are found, males were found to prefer intermediate-sized females. The hairy legs in this case made a little difference but bums were what it was all about. Males though, did not waste time on females with really big bottoms, who in this instance may have slightly cheated by pretending that they have a better than average reproductive ability. But cheating by females is not as studied as male cheating (it is not as common).

Now, although this family is called dance flies, they are not the only flies that dance. The adults of *Lipara lucens*, in the Chloropidae family rock their whole bodies while on their 'tippi-tarsus' to vibrate whatever surface they are on – they get their groove on! This was first observed in virgin females and it transpired that these females did so in response to males who were already dancing. But the female didn't need to see the male; she just had to feel his tremors. In 1981, Japanese entomologist Kenkichi Kanmiya, was first able to determine that these moves came from muscles within the thorax, and that these were stimulated by sense organs at the base of the mid-coxa in both females and males. Kanmiya later showed that these flies even had regional styles! Looking at flies in Latvia, Bulgaria, Hungary, the Czech Republic, Germany, Belgium and the UK, he found distinct differences in the way they tapped their feet – showing different 'burst' duration, i.e. times between different bursts and the ratio of the number of pulses in a burst-over-length-of-burst called the burst periods. I have images of female flies going abroad for the holidays and falling for a local male and his exotic moves.

Hairy legs, you may be unsurprised to find out, are diagnostic, i.e. they are unique identifiers. The position, whether they are on the front, side or behind, which leg they are on, the length, number and arrangement are

The sex comb of *Drosophila melanogaster*, which provides grip for the males during copulation.

all used for taxonomic identifications. Males in the subgenus *Drosophila (Sophophora)* are identifiable by a recently evolved 'sex comb' on their legs, which has developed from hairs that are found in both sexes, originally used to clean their little faces. The 'teeth' of these combs are different from regular hairs in that they orientate away from the legs and are much blunter, curvier and thicker, as well as being hardened.

To understand the function of these combs, Robert Cook, at La Trobe University in Australia, in 1977 conducted a rather brutal experiment. He 'amputated' (as he phrased it but shaved is the more appropriate term) the sex combs of three strains of *D. melanogaster* and the species *D. simulans,* and by observing their subsequent behaviour he concluded that these combs were used during copulation to anchor the male to the female – their removal did not halt copulation, but they did severely delay it. In another one of his gruesome experiments, he decapitated females to see whether comb-less males would still want to attempt to copulate with them – they did. Decapitation does not stop a fly, and many can carry on for ages – the only problem being that they eventually starve to death or suffer at the hands of a now unseen enemy.

Legs are not just for walking, but they are also very useful for tasting. It is much more efficient to run around and let your legs do the tasting than to drag your mouthparts across the surface of every substrate that you land on. And different parts of the body may be specialized for different tastes and also different reactions. By genetically blocking the neurons from these taste receptors, the gustatory sensilla, Vladimiros Thoma and colleagues showed in 2016, once more with *Drosophila,* that these receptors were critical for detecting sugar (those flies with their leg neurons blocked did not choose any sugar). Not only did they find this out, but also that there were two groups of neurons associated with the leg receptors, one that went to the brain and the other to the ventral nerve cord. The signals that went to the brain told the fly to stick its mouth out and get ready for feeding, whilst those that went to the ventral cord told the fly to stop – it was good where it was.

Legs are not just for walking or tasting, they also help with flying. The tracheae – the tubes through which air passes through the body is strengthened by rings of fibres called taenidia (how much strengthening varying throughout the body). There may be many tightly packed rings, in which case the tracheae may be very rigid, or they may be completely absent, enabling the tracheae to expand into air sacs. These air sacs are located throughout the body and are very useful for storing and delivering oxygen in areas of greater demand. But there are other reasons to have them. And that is to make the flies lighter!

A phantom crane fly, *Bittacomorpha clavipes,* found in the USA is one such creature that employs this ingenious trick. These flies have air bags in their feet that they use to float around the forest. Strictly

Bittacomorpha clavipes with its brightly banded legs and its swollen basitarsi.

speaking, it is not a crane fly but is the common name given to any species of fly in the Ptychopteridae (the phantom crane flies). Their ghostly name derives from the fact that their legs are mostly black except for an obvious white tip, and when they bob about in the shade all but these white marks seem to disappear.

When flying it appears to be floating in mid-air, with its patterned legs spread far apart. These long-suspended structures have trachea-rich tarsi. The basitarsi – the first tarsal segments – are greatly enlarged in both sexes, which can be most obvious due to the colour change in the legs (the leg parts either side are white). Charles Brues, wrote about these in 1900, and describes how he cut open the legs and then mounted them either whole or in halves. From these preparations he determined that the base part of the leg, that closest to the body, was similar to other flies. The tracheae began to enlarge as they reach the middle of the femur and then keep expanding till they were nearly the entire diameter of the femur at the end. The entire tibia was completely filled with tracheae as was the first, second and third tarsi, but the walls were not rigid as the strengthening rings had mostly disappeared. These attractive beasts have floatation devices in their feet to help them bob around the forest.

At the end of the final part of the foot, connected to the final, terminal tarsus, is the acropod, the fly equivalent of a toe. Although a distinct structure, the acropod is morphologically associated with the final tarsal segment and so is also called the pretarsus (with some authors pointing out it's actually post-tarsus). Across the insects, this region varies in the level of sclerotization, with some species opting for little sclerotization and so increased flexibility, while in others it's hardened to form rigid sclerites, or plates. If there is a claw attached to the acropod it is then attached by the unguitractor tendon, which pulls it back towards the body of the fly, the name deriving from Latin 'unguis' meaning claws and 'tractus' meaning pull.

The acropod is composed of several smaller structures: the empodium, the pulvilli and, as mentioned, sometimes claws. The empodium is a useful diagnostic characteristic and may either be pad-like (pulvilliform), or bristle-like (setiform) and is thought to act either

The acropod of a mosquito.

like the claws or the pulvilli in maintaining adhesion to a surface, as well as having a sensory function.

Literally meaning 'little cushions', the pulvilli are adhesive, flap-like wonders of the acropod. The ability of flies to land on walls, and then walk around upside down on ceilings and so on is largely thanks to them. English botanist Tuffen West back in 1862 started his comprehensive account of fly feet with the sentence, 'the structure and action of the fly's foot have been so frequently treated of, and are so generally considered to be fully understood, that it may appear, at the first glance, as if nothing further could be done with so hackneyed a subject'. Oh, was he to be proven wrong! There has been much debate since his time as to whether the mystical binding properties of the feet could be due to micro-suckers, sticking fluids or electrostatic forces.

There are two distinct methods of attachment that have evolved in insects – smooth pads, or the hairy (setose) surfaces, the latter are found on flies' feet. Both are able to maximize substrate contact, as they are both flexible, and both features have been found in the legs of various insects. Intriguingly, both structures have evolved independently of each other several times – where there is a will to stick, there is a way.

Over the past 300 years, we have been studying 'hairy' attachment systems – yes that is the actual name for them in flies – and we are beginning to piece together this puzzle. On each of the pulvilli are many tentacular extensions called tenent setae. These are not true setae (not hairs or bristles), but outgrowths of thinly sclerotized hollow projections, full of a gooey secretion, some of which have small pores at the end. Stanislav Gorb, of the University of Kiel, who has published extensively on animals and plant bioadhesion – wherein either natural or synthetic materials adhere to a surface – recently determined that the setae on the edge of the pulvilli were different in shape to those nearest the body. The nearer setae had oval plates at the end that looked like spoons, while the ones at the distal end were ellipsoid, like the end of golf clubs. The number of setae varies across species, for instance – there are 4,000–6,000 in *Calliphora vicina* but only 740–920 in *Episyrphus balteatus*.

In addition to these setae, the pad also secretes adhesive. Elisabeth Bauchhenss, a German biologist and chemist, known more for her work with spiders than flies, published one of the first papers on these secretions in 1979. Her studies of the blow fly that was then called *Calliphora erythrocephala* (now a synonym of *Calliphora vicina)* found that they were stored in a spongy layer near the edge of the pad and all insects have these adhesive secretions, whereas other wall-walkers, such as spiders and geckos, rely solely instead on what are called Van der Waals interactions. Named after the Dutch scientist, Johannes Diderik van der Waals, these are weak forces of attraction between atoms and molecules – things have to be very close for them to stick together and so geckos for example have loads of tiny hairs with cup-like processes on their ends to maximize contact.

Flies stick by the same basic principles of having loads of hairs, but also have the addition of that adhesive fluid. This fluid comprises sugars and oil, which together form a glue-like substance. This gluey stuff is left behind on the surfaces that flies walk across – they leave tiny oily footprints in their wake. The acropod is the first part to make contact, after which the pulvilli are placed down flat. To move forward they have to detach their foot from the surface, which involves four successional movements – shifting, twisting, rotating and then pulling, which allows the fly to pull its feet free from the surface and walk around. Flies can attach and detach to many a surface thanks to this – they even can attach and detach from spiders' webs.

On floors and walls, flies employ the traditional tripod gait. However, when they dangle from the ceiling, things become trickier. Have a look at a fly when it is upside down and you will see the legs are all splayed centrally around the body, a position in which all the setae are able to create the greatest contact with the surface. They are able to cling on with just two legs if they are held in this position but with one it's virtually impossible and they peel off quite quickly. The more feet on the surface, the more secure the fly is, and so they move from a three to a four feet system.

Stanislav Gorb found that the four feet attached upside-down gait pattern adopted by flies provides the best relationship between contact with the substrate and body weight of the fly. I have personally experienced a bioinspired gripping product having been 'hung' from the ceiling wearing gloves that have mimicked the feet of beetles. Unlike my insect counterparts my ratio of body weight and contact area was not that successful and I was only able to dangle for a short while and there was no way that I could move (listen to BBC Radio Four Series *Who's the Pest* to hear this event). In studying the mechanics of the fly's gait, as well as the properties of the foot, minute grippers called microgrippers are being designed, devices small enough to grasp microscopic objects.

As well as adhesive pads, flies can use their claws to grip onto surfaces (and help them off them). And there are some fantastic claws. Once more, these vary enormously, having evolved to suit the myriad different life styles that these little creatures have adopted. One of my favourite clawed families are the Braulidae – commonly called the bee lice, but I like to think of them as the bee-riders.

Oh, I adore these flies. No wings, no halteres, no scutellum, no ocelli and minute eyes. They have lost so many of the obvious morphological

A *Braula* sp. clinging to the back of *Apis mellifera*, from a hive in Warwick, UK.

features and have long caused a taxonomic headache. They were first described in 1818 by the German zoologist Christian Ludwig Nitzch. He thought they were pupiparous, i.e. they give birth to mature larvae. They are not, and so have been moved around the Schizophora and are currently placed in the superfamily Ephydroidea, as a sister family to the Drosophilids. Not only is there confusion but there are not many species – there are just eight in two genera. They reside in honey bee hives, the adults mostly found on the bees themselves, feeding on secretions from their mouths, though they can be found separate to their hosts. And boy, can these little monsters move. These have been observed to move at great speed from one bee to another and whilst circumnavigating an individual bee, a speed that can't be explained by the pulvilli-method of attachment because there are not enough setae. Their pulvilli are very small, in fact too small to sustain walking on. Instead, adults have developed special tarsal claws, which look more like combs, to enable them to cling on and run around on the backs of bees as well on the ground.

These claws have a single row of about 30 evenly spaced 'teeth', with a gap in the middle of the claw, which resembles a comb with a few broken teeth in the middle. It is these teeth that the bee-riders use to grab onto the branched hairs that are seen with honey bees. They spend most of their time gripping on to the back of the queen bee, and once the fly has mated, they will generally stay put. They need to ensure they grip on with all their might as they are wingless. If they become dislodged it's hard for them to secure a place on another host as this may have been when the queen was flying around. With the pulvilli less important for attachment in braulids they may now be used for food detection – and so they have been modified to become a sensory apparatus. This may aid in both food acquisition and something a little more interesting. Mostly only one braulid is found per bee but numbers of up to 30–40 have been recorded on a single queen. Why she puts up with this many little bodies running around her own was not known until work by Stephen Martin and Joe Bayfield, of the University of Salford, in 2014. Martin and Bayfield published their preliminary findings on the hydrocarbons found on the cuticles of braulids. Lo and behold, those masters of movement had also perfected their camouflage – the hydrocarbon profile matches that of their hosts. This very close match enables them to blend in with their colony (each colony is different), and, Martin and Bayfield propose, this has been made possible by odour acquisition. They are dousing themselves in bee 'eau de parfum'.

Many flies that live on other species, generally known as ectoparasites, have claws to hold on to their hosts. Three out of the four families of the Hippoboscoidea superfamily are ectoparasitic: the Nycteribiidae and Streblidae, the bat flies and the Hippoboscidae, the louse flies. Of the 200 or so species, many have wings, but some do not. What all of them share though is a firm grip. The swift louse fly, *Crataerina pallida*, is one such species. Adults feed on the blood of both swift nestlings and the highly mobile grown-ups. They need to consume approximately 25 mg of blood every five days, and so need to stay near their food. They exhibit several tarsal modifications in comparison to other ectoparasitic calyptrates, which includes the house fly and the dung flies. Most Calyptrate flies have single-tipped claws, whereas *C. pallida* have awesome looking triple-tipped claws, with a large blade-like heel. These claws are so good at holding on to the feathers that it's nigh on impossible to detach the fly without damaging either the fly's legs or the bird's feathers. The pulvilli are also well developed, with a high density of setae that split at their end, thereby increasing their surface area, and helping the fly to grip to smooth surfaces when necessary.

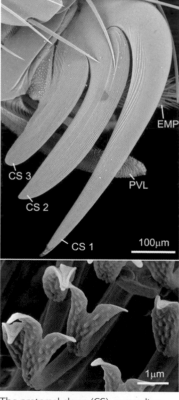

The empodium is another appendage at the end of the foot, which extends between the pad-like pulvilli. But whereas the pulvilli are for attachment, the empodium is for cleaning, according to Dennis Petersen and his team at the University of Kiel, in their 2018 paper on the subject. The empodium allows the fly to get rid of the residues of broken off feathers or setae that have clumped together. After all, if the fly's feet get clogged up with dirt then they could lose their grip, or worse.

The Deuterophlebiidae, aka the mountain midges, live almost entirely off the ground as adults. These incredibly short-lived flies live for just hours – most, if not all of which is spent on the wing,

The pretarsal claws (CS), empodium (EMP) and pulvilli (PVL), with a high-resolution image of the setae, of *Crataerina pallida*.

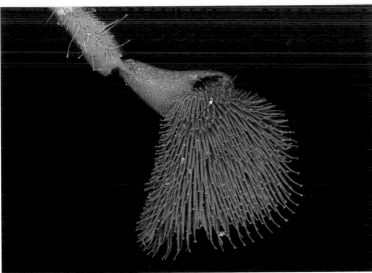

The pre-tarsal claws of the female (top) and expanded empodium of the male (bottom) Deuterophlebiidae.

flying above water. And there is but one purpose for this stage – to reproduce. And so, gone are their pulvilli. Instead, the males have just one massive modified empodium covered with tenent hairs. In most Deuterophlebiidae species, the ventral margin of the tarsus and the tibia are also covered in microtrichia – large hairs with clubbed-ends which normally help them to grip. But in the mountain midge, they may help the flies to do the opposite and stay above water. Gregory Courtney, an entomologist at Iowa State University, is one of the few who has spent a lot of time studying these odd little species, and argues that rather than helping them to grip, the mechanical and water-repellent nature of these 'hairs', enables them to exploit the surface tension of the water.

Previous work on the marine chironomids in the genus *Pontomyia*, found that they, too, had very large empodia where the males are without claws. These chironomids are part of the surface-skating groups, and so Courtney's hypothesis about the Deuterophlebiidae seems credible. The females, on the other hand, have strong claws as well as a long cone-shaped empodium. Many a female has been sampled by Courtney, minus her wings as well as a depleted amount of eggs, and has been observed clinging to the submerged rocks where the larvae and pupae are found. But why? Mated females, it transpires, drop from the swarms and commence underwater oviposition where her claws would be essential for maintaining a grip in these habitats.

Walking, floating, grooming and flirting – legs are useful for many things. Dolichopodids, along with their cousins the empidids, are the most notorious flirts. A species from Thailand, *Thinophilus spinatoides,* which was described in 2017 by Abdulloh Samoh (Princess Maha Chakri Sirindhorn Natural History Museum) and Patrick Grootaert (Royal Belgian Institute of Natural Sciences), has incredibly long legs and again conspicuous markings on its feet – multi-coloured leg warmers to attract the females. Not only are markings important for attracting the ladies, but their long legs are very helpful for being able to spring from the surface of a water body. Very long legs aid take-off by spreading the propulsive forces over a much larger area of leg and, by having larger tarsal segments, they are able widely distribute the propulsive forces and so not dunk themselves as they leave.

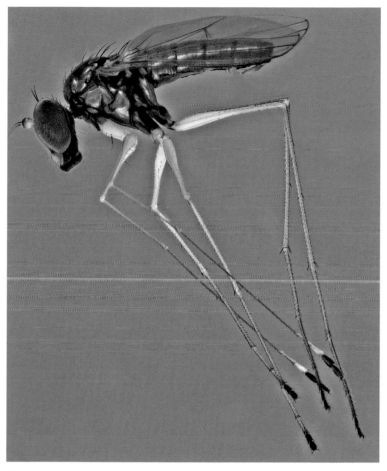

Thinophilus spinatoides (Dolichopodidae) with its long legs and coloured tarsal segments.

Another species of Dolichopodid that I often see in my garden pond is the semaphore fly *Poecilobothrus nobilitatus*, which use their hairy feet as part of their seduction routine. The male approaches a female from behind and then dangles alternate feet over her eyes. Another species found in the UK is the fancy-legged fly, *Campsicnemus magius*.

Calotarsa insignis, a flat-footed fly from a trail in Jackson County, Oregon, USA.

And it truly is worthy of its name. Oh, these males have the best legs of any animal with adornments, making them quite the dandy about the saltmarshes. Although originally described as *Medeterus magius*, this species was moved into the genus *Campsicnemus*, and this genus name is an apt one, as it derives from the Greek 'kampsis' for curve and 'knimi' for leg, describing very well the mid tibia on the males of many species. The fore tibia and tarsal segments are even more interesting, as these are ornate, with some visibly arresting protrusions of bristles, hairs and ribbons all likely for attracting and securing a mate.

When Hermann Loew, a German dipterist, described the fancy-legged fly in 1845, it caused quite a stir. He was accused by his fellow dipterist Carl Gerstaecker, of describing a species that had been deformed by fungus, because the growths on the legs looked so unbelievable. But dolichopodids have adornments all over their bodies. The Belgian Patrick Grootaert, who has described many species in this family, wrote in 2003 that in the genus *Cymatopus* alone, 33 primary and secondary sexual characteristics can be found, many of which are on the legs – such as hairy paddles and knobbly extensions.

And then you have the tibia of the midleg of the male *Rhamphomyia scaurissima*, from the Empididae family. We have come across this genus already – those devious dance flies and their silk balloons. *Rhamphomyia* are known for their amorous and often cheeky ways. There is a very peculiar outgrowth on the male's midleg, a large swelling that, from a distance, looks like a small insect – one that he has 'caught'. It is in fact a lure, to attract a female. Sneaky little so-and-so. In the mating game, one sex often exploits a need or weakness in the opposite one, in an attempt to gain the upper hand and so increase their reproductive success.

It is the genus *Calotarsa*, a flat-footed fly in the family Platypezidae, though, that boast some of the most dapper footwear. Only found in North America, and with only six species described so far, these flies, as my cousin Ros amusingly commented, look as if they are wearing gangster spats. They are the Al Capone-styled dudes of the fly world. Any fly with feet modifications that resemble such fantastic shoes have earned all the females they can handle if you ask me. All six species have a slightly different structure to these tarsal outgrowths, and so the females are able to recognise a male of her own species in a swarm. In *Calotarsa insignis* the silver part of the foot flashes in the sunlight, acting as a signal to attract female attention.

The legs, and feet of the flies are some of the most impressive structures to look at, with functions that enable diverse and sometimes ingenious ways of living. They taste, they flirt, they clean, they move, and they can produce alluring gifts – much better than me that can barely cope with just standing on mine, and I have only two.

CHAPTER 8

The abdomen

A census taker once tried to test me. I ate his liver with some
fava beans and a nice Chianti.

Hannibal Lecter

IT IS SAID THAT the best way to a man's heart is through his stomach.
That may be the case with humans, but it's the opposite in flies. It's the
female that needs to be fed if the male wants to get his lady – he needs
gifts of food to woo her as the female has to transfer the nutrients that
she has obtained from food into producing and sometimes nurturing
the next generation of little maggots. The final part of the fly's life is all
about producing the next generation, and it seems only fitting that this
happens in the final segment of the fly's body, the abdomen. It is the
part of the body specialized for digestion, defecation and fornication.
Of course, this is where we find the genitals, but I am going to leave
them from this chapter because the genitalia of flies, both sexes, are so
wonderful that they deserve a chapter all of their own.

The abdomen of flies looks like the least modified section of the
body. It lacks the showstoppers of the head and thorax – the wings,
the legs, and enough sensory equipment to sniff out a dead body from
16 km (10 miles) away. The abdomens of insects have lost all of their
appendages, unlike those of their close relatives, the millipedes and
centipedes. Millipedes, for example, have a three-segmented thorax with

Dance flies, *Empis tessellata*, mating while the female eats a gift presented by
male, in Bedfordshire, UK.

no legs on its first segment but a pair on the following two. The millipede abdomen is very similar in appearance to its thorax, with limbs coming off it, each segment (technically two segments fused together) with two pairs of legs, which they march along on. But insects, including the flies, no longer bother with legs all the way along and have dedicated the abdominal region to nicer things. Their abdomen is deceptive.

The abdomen is the most flexible part of the body (external features not included), as it needs to be able to expand and contract to accommodate food and eggs as required. Don't for a second think that because they have lost their external appendages that the abdomens of flies are all dull. Quite the contrary, there are some visually arresting abdomens out there, and some of my favourites belong to soldier flies – the Stratiomyidae. I affectionately call these the 'fat-bottomed flies', as some of them do indeed have enormous nether regions. A great example of this is *Platyna hastata*, whose abdomen is nearly as wide as its body is long! I can't think of any other animal that has a derrière of such proportions – Kim Kardashian has a famously large rear end, but hers has nothing on these species.

Platyna hastata is widespread across tropical Africa, where they can be seen congregating in large swarms. Martin Hauser, who is based at California State University, but who is just as likely to be found wandering around the planet collecting Stratiomyidae, wrote to me about how he had observed them swarming, and that generally they like doing this in a clearing with, as he beautifully describes it, 'sunlight against the dark of the forest' bobbing up and down, sometimes in a tight swarm but mostly in loose groups with their abdomens akin with 'blinking stars at night'. The big flashy bottoms and swarming behaviour can also be found in other groups of African stratiomyids, including the related groups of *Platynomorpha*, *Platynomyia* and *Enypnium*.

The Stratiomyidae family are not the only ones with quite literally flashy bottoms. In the Milichiidae family, often known as the jackal flies, there are many shiny or conspicuous-bottomed males – for example, *Milichia patrizii*, who swarm for mating, their silvery abdomens visible as a lure to females from a distance. The male *Milichia patrizii* has a very white bottom, and forms mating swarms, whilst the female's abdomen is as dark as the rest of her body – she has no need for such ostentatious outfits.

The length of the abdomen varies more than the width. The nematocerous flies, such as mosquitoes and midges, and the Brachyceran species, such as the soldier flies, have, on average, nine segments. The rest of the higher diptera have reduced the number further, and on average the males have five and the females six segments.

Platyna hastata (Stratiomyidae) and its abdomen that is almost as wide as the body is long!

The genitals sit at the end of the abdomen, followed by the proctiger – the weakly sclerotized (hardened) terminal segment that bears both the anus and a pair of sensory organs called cerci. The cerci are packed with sensilla that are thought to help with copulation – by 'guiding in the bits' as it were, as this has been observed in other animals that have similar appendages. Not all adult males have them though, and they vary in length and number of segments, with the more primitive nematocerous flies having well-developed, elongated and two-segmented cerci, which become reduced in the higher flies to smaller sub-spherical or oblong

The beautifully sculptured ovipositor of a female *Tanyptera* sp. (Tipulidae).

structures often composed of only one segment. Some females have gone one step further and hardened these structures and fused them together to form a long tube from which they deposit eggs: an ovipositor. You see this type in the fruit- and vegetable-piercing groups, such as the tephritoids, the fruit fly families, and the tipulids, the crane flies.

None of the abdominal segments are as sclerotized as the head and thoracic regions though and this is to enable them to expand and contract when the fly is feeding or gestating. The sclerotized plates are organized into two rows down the length of the abdomen: the dorsal plates or tergites along the upper side, and the ventral plates or sternites along the underside. The tergites are toughened more than the sternites, and the two regions are joined by a membrane, again to allow for growth and subsequent retraction.

Inside all this is the region where most of the digestive processes occur. And this digestive or gut system is remarkable. Some species are able to expand themselves to extraordinary proportions as they feed, and this is especially prevalent in the blood feeding species that have to take in a large meal, quickly. There is something very attractive about the abdomen of an engorged tsetse fly. The larger tracheae can now be clearly seen as large ribbon-like structures that wrap the abdomen,

An engorged tsetse fly (Glossinidae) showing the tracheae around the abdomen.

causing it to look like radicchio, the Italian chicory. The stable fly *Stomoxys calcitrans*, for example, takes between two to five minutes to take in enough blood to enable egg development to start.

The digestive tract of flies has to cope with a variety of food sources. Often the same individual is feeding on both blood and nectar, so the interior has been divided into a series of distinct regions, for consuming and processing meals, then for nutrient absorption, and finally for excretion or waste removal, the length and size of the different parts varying with species and food type.

The gut of an adult fly is an enclosed tube that runs from mouth to anus. The foregut runs through the head and thorax, and on to the mid- and hindguts that are situated in the abdomen. The food is broken down by both mechanical processes and digestive juices, and sometimes stored, in the foregut. But it is in the mid gut – the mesenteron – where most of the digestion takes place.

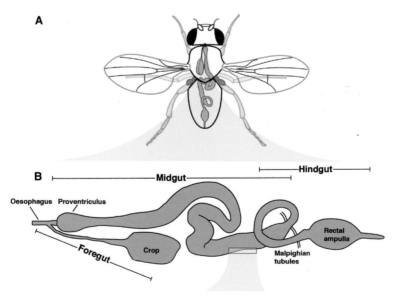

A

B |————————— **Midgut** —————————| |————**Hindgut**————|

Oesophagus Proventriculus

Foregut Crop

Rectal ampulla

Malpighian tubules

The gut of an adult *Drosophila melanogaster*.

The mid gut has been poorly studied, which is quite remarkable considering how essential a good diet is to the fly. Flies, like us, have a variety of micro-organisms that live inside them to aid digestion. It has been suggested that the number of these microorganisms could be greater than the number of cells in their entire body – a great pub quiz fact if indeed it proves correct, and it is more than likely to be considering that they are very small (much, much smaller than the fly's own cells). The larvae of the mosquito *Aedes aegypti* don't start off with any gut fauna and the poor things can't actually grow if they are prevented from establishing a healthy microbe community, cultured from microbes they would have consumed from their environments.

The midgut bacteria have been shown to strengthen the immune response in mosquitoes and protect them against any unwelcome pathogens – and yes, although some flies are the spreaders of disease, they too can fall victim. The bacteria in the gut of adult *Anopheles* mosquitoes inhibits *Plasmodium* infection – it alerts the immune system of the mosquito to let it know that these parasites have arrived and that the host needs to do something about it, but these bacteria are also directly attacking the *Plasmodium* by producing enzymes and toxins. There is a complete battle going on in the guts of these tiny mortals. Studying which bacteria are the most effective at tackling this will aid our war against malaria. As mosquitoes are smaller, and less complicated than us humans, they also make a great model for understanding the interactions between the host, the good bacteria and the bad pathogens.

Not only do these bacteria keep pathogens at bay but they are also essential for egg development. My good friend, Dr Iain Haysom, a senior lecturer in microbiology at Bath Spa University, has been screaming to me for years (we undertook our PhDs at the same time) to recognize that the minuscule bugs are as important as the bigger 'bugs', and of course he is correct. How important? Analiz de Olivera Gaio and co-authors (all based in Brazil) in 2011 showed how critical a contribution the bacterial gut flora makes in blood digestion in *Aedes aegypti*. The team fed females different antibiotics to kill off the microbiota of the gut. It transpired that the gut fauna was critical for breaking down the ingested red blood cells of the mosquito's meal. They found that most of the bacteria identified undertook some of this so-called haemolytic activity, with *Enterobacter* sp. and *Serratia* sp. being the most active microbes. The bacteria are essential for rupturing the red blood cells, and a reduction in the bacteria led to a reduced uptake of blood proteins from the meal, ultimately reducing egg production.

Diet among flies, as with everything else, is incredibly varied. Dr Victor Brugman, now at the London School of Hygiene and Tropical Medicine, undertook the largest study looking at the blood meals of mosquitoes for his doctorate. From nearly a thousand individuals from 11 species, Brugman found that there were 19 different hosts, 14 of which were birds and the remainder mammals. This is interesting in terms of conservation as we have suffered from a series of penguin deaths in our zoos in the UK, including London Zoo, where in 2012 six Humboldt penguins *Spheniscus humboldti* died. These birds are contracting avian malaria, which is spread by the mosquito genus *Culex* rather than the human malaria vector *Anopheles*. Antimalarials are now given to the birds.

There are more than 60 species of malaria parasites and how fussy the mosquitoes are in feeding will affect the transmission of these species. Humboldt penguins are native to the coasts of Chile and Peru, where mosquitoes are kept at bay because of the cold temperatures. London is a very different habitat, where migrating wildfowl can pass on malaria to our resident mosquitoes and infect species that have no natural immunity to the parasites. This is not just a worry for captive populations of birds but also for those native populations; due to changes brought about by human induced climate change, warmer weather is likely to allow the mosquitoes to flourish. The composition of the gut microflora is related to the host type and, in a changing world, this has implications for all vectors as some will be able to adapt better in new environments than others.

Now, the processing of these blood meals in the fly's digestive system takes time. The female may take several days to assimilate it, and she is often feeding on nectar as well, both of which she may ingest in large volumes. Cleverly, the sugar meal is stored in an upper part of the gut called the crop, which can be sectioned off from the main alimentary canal. The sugar is then released in small amounts when required, so that her stomach largely remains empty in constant readiness for a blood meal. Even more amazingly, when she has blood in her stomach, the stomach's lining secretes a semi-permeable layer called a peritrophic membrane, which gloves the blood meal and keeps it separate from the sugar solution.

The blood is solely for the synthesis of egg yolk proteins, which come from the breakdown of the ingested blood into its constituent amino acids – the building blocks of proteins. Once a blood meal enters the gut, trypsin – the enzyme that breaks down the proteins – is released and may take a long time to be fully released. For example, in *Anopheles stephensi* it will continue to be released for up to 60 hours – that is, two and a half days later. But even with a bubble of blood in her stomach, the female can still take up the nutrients she needs to survive (the blood is just for her offspring). Females can double or triple their body weight, and so she reduces her liquid diet to mostly solid nutrients by excreting most of the excess fluid (as well as excess salts), and often does so while feeding.

As well as being the region where most of the absorption occurs, the guts also remove the waste products of the haemolymph – the fly's 'blood'. Mosquitoes have kidneys called malpighian tubules (MTs), named after Italian scientist Marcello Malpighi. Malpighi is known as the father of histology – the study of microscopic tissues. He had a family of plants named after him, he was the first to observe trachea in insects, and he studied the human brain and, as we've seen, the excretory system of insects. Malpighian tubules arise from the hind gut and absorb solutes, water and waste from the haemolymph. This can be excreted as a solid or, in the case of the female mosquitoes and other blood-feeding species, as a lot of liquid.

The MTs are just swashing around in the haemolymph, so for waste products to pass through the walls of these tubules, there are pumps that transfer potassium ions from the blood into the cavity. Hot on their tails are the negatively charged ions to maintain electroneutrality which, in turn, thanks to something called the 'osmotic pressure gradient' (meaning the water will flow into a more concentrated solution) is followed by water and all the crap that the fly wants to get rid of, e.g. salts, sugars and urate ions. This filtrate, called the primary urine, gets mixed in the mid gut with all the solid parts of the meal and soluble products of digestion and passed on to the rectum. Flies don't pee urine like mammals, but instead convert it into insoluble uric acid which, depending on how much water is reabsorbed, can come out with excess fluid or as solid little pellets or the tiniest of poos.

The bioluminescent larvae of *Arachnocampa luminosa* – the light being emitted from the end of the Malpighian Tubules.

These MTs are essential for osmoregulation and excretion, and they are also responsible for creating one of the most visually arresting underground sites in the world – that of the bioluminescent larvae of the New Zealand glowworm *Arachnocampa luminosa* (Keroplatidae) which can be found dangling from ceilings of the Waitomo cave in New Zealand in a display of deadly beauty. For it is the swollen tips of the end of the MTs that create these bioluminescent structures to lure prey to their deaths, as they get caught on the sticky strains of silk that the larvae use as fishing lines. They are the only animals where the MTs have been developed for this. The enzyme that

A mosquito with its bum bubble – a clever way to avoid thermal stress by evaporative cooling.

produces this bioluminescence is from the luciferase group, but in this species the luciferin compound that the enzyme works on to produce the light is completely different to those used by other glowing creatures. Flies are so brilliantly unique.

There are further issues to overcome when feeding. Mosquitoes and other blood-feeding species have to cope with the high internal temperatures of their hosts, which could have a deleterious effect on them. To limit this, many blood-feeding species are heterothermic, i.e. they vary between self-regulating their temperature and using the external environment to regulate it. The *Anopheles* genus does this by excreting fluid droplets from the end of their abdomens. This reduces the temperature of the internal fluid by a process of evaporative cooling, where the external air cools the fluid. This mechanism prevents all the gut symbionts, as well as the mosquito, from overheating.

So, as you see, not everything that comes out of the backside of a fly is bad. The flies, as we know, are the connoisseurs of romantic gestures, and what is more romantic than an anal secretion? *Drosophila grimshawi,* one of the 600 or so species of Drosophilidae that live in

Hawaii, has optimized its anal secretions for maximum effect. The males drag their abdomens over their perches, such as leaves or branches, and thus streak pheromones across the surface. The more streaking, the more success at attracting the opposite sex these little males achieve – though I can't see this technique having the same effect for males of *Homo sapiens*.

Anal secretions are also the domain of *Ceratitis capitata*, the Mediterranean fruit fly or medfly, from the fruit fly family Tephritidae. This stunning little creature brings much heartache to many, as it is one of the most destructive fruit pests on the planet, with the larvae consuming more than 260 different types of fruits, flowers, nuts and vegetables. Males defend individual leaves to act as a stage on which they compete for the attention of the opposite sex. All the while, they secrete a pheromone from their anal glands, specifically from everted rectal epithelium, in the form of glistening bubbles. These pheromones are not simple chemical compounds. To date, 56 different compounds have been identified in the odours that surround the male *C. capitata*, of which five dominate. Not content just to rely on his own smells, he also makes use of the plant's volatiles – that is the plant's odours. These odours attract the female to the general area as she searches for a suitable larval habitat.

Flies also use anal odours to defend themselves. Yes, they fart to warn off potential attackers. Take species in the family Sepsidae – the scavenger flies, for instance, which resemble flying ants in their appearance. Sepsids are the only known family of diptera that have a defensive abdominal gland called the Dufour gland, named after Léon Jean Marie Dufour, a French medical doctor and naturalist, who described the gland in 1841. This gland is found in all of the Apocrita, the suborder of Hymenoptera that includes the bees, wasps and ants, and has a whole host of purposes – mostly for communication but also for lubrication of the ovipositor and food production for the larvae. The sepsids use it for communication as well. Certain species of sepsids produce a sweet odour composed of a mixture of solvents and defensive chemicals. The wonderful British dipterist, Adrian Pont, in a paper published in 1987, describes how the smell from swarms of these flies has the effect of 'stupefying large numbers of individuals' (flies not humans). These are huge swarms that, instead of lasting for a couple of hours, have been recorded persisting for months, tens of thousands of flies thick, remaining in exactly the same location. And apparently, they smell of honey – though Pont likens it more to the smell of a cockroach. The chemicals that produce this odour will have been secreted, via the anus, from scent glands on the dorsal wall of the rectum. What is unusual is that both males and females secrete it, and these secretions are thought to be for both protection and aggregation signals – marking and memorizing meeting places for example. Pont noted that if an individual in one of these mass swarms was say, attacked by a spider or a wasp, they were often released unharmed, their odour being a repellent to the predator.

The mass swarms of this species are not thought to be mating swarms seen in many of the other dipterans, but rather as a means to call everyone together before hibernation. The secretions are thought to be general aggregation signals, as they mark the places where they can meet again the following year for mating. It's a bit like a lover's note to tell them where to meet the following year for their holiday romance.

Sepsids are a great bunch of flies whose reproductive behaviour is incredibly diverse, so we use them as models for sexual selection research. Much of the recent work on this family comes from a laboratory at the National University of Singapore under the auspicious leadership of Professor Rudolf Meier. Back in 2014, 400 or so Dipterists, including myself, converged on Potsdam in Germany for the International Congress of Dipterology, and were treated to a key lecture on the mating behaviour of this family. Meier and his students showed a series of short films they had recorded that documented this amorous family. I was doing fine until one of the videos showed the flies kissing. The male was recorded kissing the top of her head and thorax – I felt a bit like a voyeur. You, too, can watch these videos, as they were published in 2009 in an online paper called *From Kissing to Belly Stridulation*. As well as kissing at the front end of their bodies, there was also 'heavy petting' at the other end.

One of my favourite things about some of the males of Sepsidae is a development on the abdomen, what Kathy Feng-Yi Su, a student in

A male *Themira superba* with his superb sex bristles.

Meier's lab, describes as a 'sex tickler', in a 2017 paper she co-authored. These clusters of hairs on the underside plates of the abdomen have the sole purpose of stimulating the female's abdomen during the mating process. These lower plates, the sternites, can range from being very simple to what can only be described as being the most exuberant of appendages, ones that can even move.

A species that has taken these lower abdominal brushes to the edge of good taste is *Themira superba*, identified by Alexander Henry Haliday, an Irish entomologist. He first described this new species in 1833, from his local area of Holywood, County Down in Ireland, and notes in his species description that it was a 'very distinct species, with twisted and spiny front legs, and tufts of hairs on the hypopygium (a modified abdominal segment in which the copulatory structures arise) that are as long as the abdomen. I think that I can agree with that description. That the male's bristles can equal the length of its abdomen is most impressive – indeed, superb. These features are a good example of structures that have evolved entirely under sexual selection pressures.

Not to be outdone by *Themira's* weird abdominal structures, some Dolichopodidae have developed their own unique features – a cingulum – that emerges between sternites 4 and 5. This odd erection comprises two appendages called signa, that are displayed aloft to look like ribbons or flags. The cingulum is most developed in the genus *Scellus*, whose 25 species live throughout the Palaearctic and Nearctic regions of the northern hemisphere and are generally larger than the average Dolichopodid. They frequent salt or alkali flats, and although we are not totally sure of the purpose of these appendages, it's thought they might be a form of visual signalling and/or, due to the tufts of hairs, a dispersal mechanism for pheromones in the males.

Odd abdominal structures that act as lures or produce smells are not confined to males, as dorsal abdominal glands (DAGs) are also found in females. *Megaselia* (Phoridae) is an enormous genus of more than 1,600 species described to date. Brian Brown, a global authority on this group, co-authored a paper in 2015 with Wendy Porras, both at the Natural

History Museum of Los Angeles Country, to describe extravagant female sexual displays in an as-yet undescribed species, of which there are possibly another 20,000–30,000. The presence of these glands has been known for many years. Henry Disney, a UK specialist on Phoridae, published a paper in 2003 solely on Phoridae DAGs and in fact used these structures to establish a higher taxonomy for them, as they are solely found in

The adult females of *Vestigipoda* (Phoridae) in the nest of *Aenictus* ants.

Does my bum look big in this? Let's hope so. A female *Megaselia* species displaying her sacs to an approaching male.

the subfamily Metopininae (although that is an enormous subfamily as it contains *Megaselia*). Disney split the tribes in the subfamily by the position of the glands on their abdomen. And it was the presence of these glands that enabled him to work out where a truly remarkable genus of phorids, called *Vestigipoda*, fitted in. The adult females of this genus (of which only five species have been described) live in the nests of ants and not only do they have vestigial wings, but they also have vestigial legs (hence the name) and so look just like ant larvae. The females also have a fan-shaped wick, what Disney believes has evolved from a mutant bristle that is believed to release the ant-mimicking chemical.

The dorsal abdominal glands lie on top of the abdomen, either between tergites 4 and 5, or 5 and 6. What is also rather wonderful about these species is that some of them have eversible sacs associated with them, those which can be turned inside out. Brown's paper has some wonderful images of these sacs doing exactly that, and as they do resembling car air bags. It is thought that these sacs make the female appear larger and able to have more offspring, as it increases the size of her abdomen, as well as emitting pheromones (and as such are very

A female *Phalacrotophora fasciata* with her eversible sacs fully displayed and the male with his head firmly attached to her derrière.

similar to the hair pencils and coremata - the pheromone producing structures seen in male moths).

The most impressive DAGs I have come across are in *Phalacrotophora fasciata*, another Phoridae. Females have an extraordinary pair of eversible sacs that when inflated look like the beefiest of handlebar moustaches. The male's head in this instance becomes encircled by the sacs, which focuses the lower part of his body firmly at the abdominal end of hers, presumably drenching him in her pheromones. She uses her wings to aid the dispersal of these pheromones and, Brown suggests, these may spread over long distances.

Bristles are a persistent feature on the abdomen, and one of the bristliest families is the Tachinidae. More than 8,500 species in this family have been described which, although morphologically diverse are, with only a few exceptions, all extremely bristly. Most are so distinctly bristly that you can easily recognize them as a family while out and about, without the aid of a microscope.

One of the many spikey-bottomed species, or as I have heard them called, bristle-butt flies, are in the genus *Anacamptomyia*, another

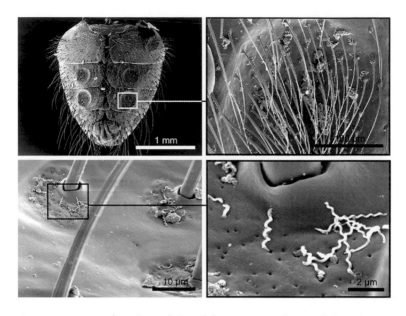

Anacamptomyia sp. from Senegal. Top, abdomen in ventral view, whole and close-up of a sexual patch; and bottom, the same sexual patch in ventral view and in close-up on tergite four showing the waxy coils.

group of tachinids. As well as bristles across the abdomen they, along with many more genera, have especially hairy regions known as sexual patches or sex patches. These are covered in stiff bristles as well as incredibly fine hairs called microtrichia, the density and composition of the sensory structures varying across the genus. Underneath these patches of specialized hairs, the epidermis has thickened and contains a layer of secretory cells. The number and position of these patches varies across the Tachinidae, but all have them, at least on tergite four, one of the plates along the top of the abdomen. This is the same place that the flag-like cingulum of the Dolichopodidae are found, highlighting the diversity of sexual specializations among flies.

A 2014 publication by Pierfilippo Cerretti, an Italian dipterist, and colleagues described these patches as 'external windows of a complex of exocrine glandular organs'. Not quite the windows to a fly's soul, but these do deliver what are thought to be special secretions that Cerretti speculates could either be for species aggregation, for courtship, or for marking those already mated, and that when present the carpet of bristles and microtrichia acts as a sponge for the secretion. These waxy secretions are produced from oenocytes, which are the cells found in various locations in the body of a fly. Those in close association with the epidermis that are responsible for lipid processing and detoxification, and these lipids, including the cuticular hydrocarbons, are the essential pheromones that tell other species and sexes so much about the state of the individual. These waxy secretions are smeared over the individuals own body as well as its paramour. The team did not identify the secretions they noted as residues on the abdomen (the curly-whirly shapes) but these do appear to be a waxy substance, similar to those described in the other flirty flies.

Sex pheromones are not unique to flies. More than 1,600 different molecules have been determined as sex pheromones across animal orders. But insects do dominate this group and use olfaction or smells as a major means of communication, especially to the opposite sex. *Lutzomyia longipalpis* (Psychodidae), is in the subfamily Phlebotominae, commonly called the sand flies, and this species is a primary carrier of visceral leishmaniasis, a fatal protozoan parasite that causes 20,000 to 40,000 human deaths a year. Unwittingly, the fly has been used by this parasite to spread itself among the human population. *Lutzomyia. longipalpis*, as with many vectors, resides in a species complex (*L. longipalpis* s.l.), with distinct differences in the spot patterns on the abdomen in different regions across the world. These spots are thought to be where pheromones are released. Interestingly, there is a notable lack of large hairs, or macrotrichia, in these regions, which may facilitate the chemical dispersion. These pheromones stimulate the female prior to copulation – the 'Lynx effect' of the insect world.

Much of what we know about *L. longipalpis* and its associated pheromones is due to the work of Richard Ward and collaborators, on infectious diseases. Ward spent many years in Brazil studying this species complex and was able to determine categorically that these glands act as sexual stimulators. He did so by removing the entire glands from males and smearing the extracts on to paper discs, whereupon he found that females were attracted to these discs from distances of up to 60 cm (23 in) away.

Knowledge of the pheromones and how they can attract over distances has been used by humans in pheromone traps for species that we consider pests. Synthetic pheromones have been developed for attracting both females and males of these species, and so can be used as a species-specific way of targeting vectors of both agricultural and animal diseases.

Sometimes an abdomen needs to move in order to attract attention (it works with humans, why would it not work with flies?), and some of my favourite things in the world in this regard are the bee flies, from the genus *Bombylius*. We have four species of this genus in the UK, but there are nearly 280 described species distributed around the world. The females have a unique, pouch-like structure on the final abdominal segments, called a sand chamber, which they have been seen to 'twerk' –

Bombylius filling her sand chamber, California, USA.

for those of you who have never seen this kind of movement in humans, it is dancing in a squatting, provocative fashion, waving your bottom around. These flies are not doing it for the sex appeal, but rather for the purpose of getting sand into this pouch. Females use the sand to coat their eggs, to stop them drying out, as the membrane that surrounds a bee fly egg is not waterproof. This pouch is formed ventrally by the turning inside out of sternite 8 on the underside and by tergite 8 on the abdomen's top side, and then covering it with long hairs. Females also bear spines at the tip of their abdomen that she uses to dig into the soil to aid oviposition. She twerks, takes in the sand to dust her eggs with, and then hurls or buries them near or around the nests of the solitary bees that her offspring will consume. I am lucky to have a healthy population of *Bombylius major*, the dark-edged bee fly, in my garden, and at the start of spring these little powder-puffs on legs are all over the flower beds and lawns, twerking and flirting, a sight that I will never tire of.

Shiny, spiky and smelly are the abdomens of flies. Wondrous in their form, albeit slightly messy in their nature. They are arguably the most underrated part of a fly's body and, particularly in functionality, the most under-studied.

CHAPTER 9

The terminalia

What I've got you've got to get it put it in you.

Red Hot Chili Peppers

T HE MEETING OF GAMETES, that is, the male sperm and the female egg, is what sexual reproduction is all about, creating a next generation that is a touch different from the last. All that eating, flying and running are just to ensure the flies get to this point – making little flies. We have already seen the diversity of the fly's adult body, but it's with the ever-so-critical genitals, which bring the gametes together, that they have excelled themselves. Anyone who has ever had a conversation with a Dipterist knows that they are obsessed with them (it's not just me, honest). Never have I thought that mammals were more disappointing than when I first compared their genitalia to that of flies.

Vincent Wigglesworth, the British entomologist whose work in insect physiology was significant, describes the act of reproduction thus: 'the egg cell liberated by the female will develop only after fusion with the spermatozoal cell set free by the male' – a magnificent release of gametes, which may not happen at the same time with males and females, but has led to the development of a huge range of both primary and secondary sexual characteristics. The end of the abdomen is a complex of genital and anal segments, and many other modified appendages associated with copulation and egg laying. These are collectively referred

Close-up of *Efferia aestuans* copulating, showing the abdominal twisting.

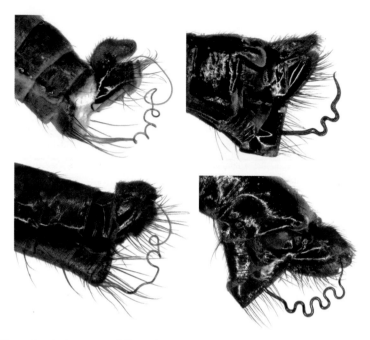

The curly-whirly penises of *Rhamphomyia* species.

to as the terminalia. Adult male flies have developed some of the most elaborate terminalia in the animal kingdom, associated either directly or indirectly with sperm transfer, as well as claspers and combs to help with copulation.

It is incredibly difficult to introduce the many features of the male genitalia in flies, as they are so wide-ranging. They have been examined in far more detail than female genitals – many of the males have large, external genitalia that are easy to examine, as many of the parts are hardened. Animals that practice internal fertilization (rather than the external fertilization as you might see with, say the mass spawning of corals), have males whose genitalia are some of the most complex and most rapidly evolving morphological structures. Back in 1844 it was proposed that males and female insects had a 'lock and key' mechanism – there was one specific male 'key' for the female's 'lock', but this is probably not the case for many species, including flies. For many female flies whose internal parts are not obvious or robust, and often remarkably alike across similar species, there were lots of different keys for just one lock. Another theory as to why there was such variation in male genitalia was perhaps due to the accumulated pleiotropic effects of genes – when one gene influences more than one morphological trait – and that the genitalia were getting the brunt of this doubling up. This hypothesis does not have much support.

The real answer is probably due to sexual selection and/or sexual conflict. There is a war going on inside the females which, until recently, was hidden to researchers, and is one which we are just beginning to figure out. Males have to out-compete other males for being first to mate with a female. If she has already mated, his sperm has to beat any other sperm already inside her to fertilize her eggs and also hopefully dissuade her from further mating events. The females are way subtler and undergo what is called 'cryptic female choice', where she chooses internally what sperm to use (but more on that later in this chapter). It is the conflict between these two opposing forces that produces such eye-watering genitalia.

The terminalia for both males and females consist mainly of genitalia, across segments six to 10 of the abdomen. The nematocerous flies have fewer segments; for example, the nematocerous male terminalia (in mosquitoes and crane flies) generally consist of segments 8 to 10, while the more advanced Cyclorrhaphan males (including hover flies and phorids) also have segments 6 and 7 to play with, as it were.

The extracted terminal parts of a male fly *Peckia* sp. (Sarcophagidae).

When you start looking at the extraordinary male genitals, understanding which bit is which can be quite overwhelming, not to mention what the different bits do. We still have much to learn about the functional morphology of many of the parts but advances in scanning and visualizing techniques has enabled some progress. Once more, we turn to *Drosophila melanogaster* for some insight into what happens both during and after copulation. A team led by Alexandra Mattei at Cornell University in 2005 was the first to show the world the internal workings of what was happening during and after sexual relations, thanks to 3D micro-computed tomography scans. Up to this point, it had been all very well dissecting out the complex gentialia but understanding how they all pieced together was not easy. Mattei and her team were able to see specific changes in the female's reproductive tract post copulation, brought about by both the male's seminal fluids and the female's secretions.

Drosophila copulate with the males positioned over the back of the females, but this stance is not standard across the order of flies. How flies position themselves to copulate is in fact very useful for understanding how species are associated with each other, and so piecing together their evolutionary relationships. The fact that, during copulation,

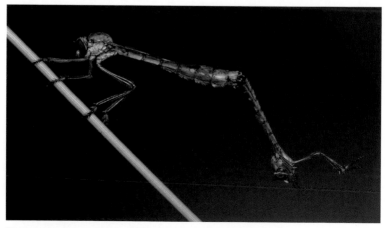

The upside-down mating of *Leptogaster cylindrica* (Asilidae).

some males are dragged behind the females or dangled upside-down with no apparent care for his welfare, while in other species the males retain some dignity and are able to mount her, are all of interest to the taxonomist (and quite frankly of amusement to me).

For flies to reproduce, both the male's and female's underneath parts need to be in contact with each other. This can involve a whole lot of twisting and flexing, which is often only made possible by some crazy asymmetry of his terminalia, be it permanent or temporary. Many of the nematocerous males have symmetrical genitalia, so he has no choice but to turn himself upside down to get the deed done. Bilateral symmetry, or something approximate to it, is the default shape for most animals, but there are of course many exceptions, including the ears of owls, the claws of fiddler crabs and human testicles. In the case of many arthropods, there is widespread asymmetry of the genitalia, where they may have moved across to one side or the paired structures are no longer the same size or shape. There are asymmetric nematocerous flies e.g. some fungus gnats and some biting midges, but oddly none have been found in any of the lower Brachycerans. Some males, including the robber flies undergo a temporary twisting, that is referred to as the 'false male-above' position, where he is sitting on her, but rather than his

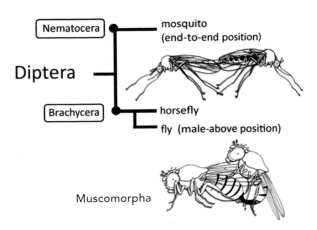

Mating positions in flies – from the males being behind to the males on top.

genitals having rotated, he has to twist his abdomen around, so he can 'connect' with the female below. In some of the Brachycerans, the genitals have gone as far as to rotate around the body, to accommodate a better mating position as well as having some asymmetrical structures.

In 2019 Momoko Inatomi and co-authors, from Osaka University, wrote about the evolutionary shift from end-to-end mating (originally called the linear position), to the male-above position (sometimes called the male vertical position). During end-to-end mating, the males get dragged backwards behind the female if she decides to move mid-act, which is far from ideal. The male-above position is infinitely more favourable for the male as he can see where he is going if the female moves and can be on the lookout for any dangers. The male-above position has been made possible by the evolution of the male's genitals, both in terms of position and symmetry of structures. Asymmetric genitalia are to be found mostly within the group we call eremoneura – the Empidoidea and the muscomorphan flies, and by golly gosh, there are some amazing examples, ranging from the rather discreet to the absolutely ludicrous. All eremoneura have a pair of articulated lobes as part of the male genitalia called surstyli (singular surstylus), which are used like a pair of forceps for the male to grab on to the female. These are not always symmetrical either – in the genus *Ocydromia*, (Hybotidae) they are lopsided, where one is larger than the other.

The *Ocydromia* is an example of a rather moderate asymmetric structure, but the closely related Dolichopodidae, infamous in dipteran circles for their massive genitalia, contain more extremely unbalanced structures. Take for example *Thinophilus langkawensis*, a newly discovered species of marine dolichopodid from southern Thailand, described in 2017 by Thai researcher Abdulloh Samoh and co-authors. This species has an extraordinarily large and rather lopsided genital capsule, the part of the abdomen that is often hinged in dolichopodids within which the male apparatus lies. The surstylus is moveable and connected to another structure used for clasping, the epandrium, and so enables the males to penetrate the female in a position more to his liking.

Even for a dolichopodid, this species is particularly well-endowed, a fact I shared with Sir David Attenborough one day when I was lucky enough to do some filming with him. He turned to the film crew after seeing such a structure to inform them that he felt quite inadequate.

Any change in mating positions in flies has required a degree of co-operative evolution between the sexes, in terms of the couple's relative position and the position of the male's genitalia. The rotation of the genitalia can be temporary (it can twist around) but more often this is permanent, occurring in either the pupal or adult stage. For example, it takes male phlebotimines (Psychodidae) 16 to 24 hours after they have emerged from the pupal stage for their bits to get into its mature position. In dolichopodids, the genital rotation is 180 degrees, but the permanent rotation in Muscomorpha is achieved after circumversion, a rotation of the whole 360 degrees from the original position (rotation of muscle that these parts are attached to is easier than of fixed body parts).

All of this rotation in diptera relates to the many different mating strategies, from the swarms that we amble underneath on our evening walks, to males clustering in leks, where the males get together to display in the same way that their much larger mammal counterparts do, to random meetings whilst on the wing. Swarms greatly aid the crepuscular species, those that feed in the twilight hours, e.g. species of mosquitoes that blood-feed at night do not have to search a lot as she knows where to go to get the sperm and lots of it, all ready for egg development. But swarms can be costly for females; they often suffer from the attention of too many males. Swarms may not be good for males, either – all that competition for the female's attention. Another less stressful method for females is

The very well-endowed male *Thinophilus langkawensis.*

to sit back and let the males fight it out first. Lekking occurs with many species of fly, and often results in some of the most ornate structures in males, such as the eye stalks and cheek processes that we saw in Chapter 2. The change mostly coincides with the males being unceremoniously dragged along by females to males mounting the females. The nematocerous and some lower Brachyceran flies tend to mate on the wing, in a swarm, whilst the more advanced muscomorphans generally mate whilst stationary and resting on a substrate – sitting down as it were. This is not always the case – Platypezidae, Lonchaeidae, Milichiidae, Fanniidae and Anthomyiidae include many species that swarm but whose genitalia has rotated. The difference with these species is that, upon coupling, they descend from the swarm to complete the act 'sitting down'.

Once these species are in cop (copulation), the process of sperm transfer begins, and one of the major changes in flies in relation to most other insects, is the development of a sperm pump, where sperm is delivered internally into the female as a fluid mess. The sperm pump is what gave the grouping of the three orders, scorpion flies, fleas and true flies their collective name – Antliophora – as 'antlia' is Greek for pump. But this characteristic is not the same for the three orders – their structures differ.

In many of the nematocerous flies, the copulatory organ, the fly equivalent to a penis, is a simple appendage, which may or may not be evertable (goes in and out). The part nearest to the body is the sperm pump that is the ejaculatory part – it comprises a sperm sac and an ejaculatory 'apodeme' – a rod-like sclerite that acts like a piston. Sperm travels from the internal testes down a tube into the sperm sac, and then *whoosh*; it is pumped at speed out through the ejaculatory duct, and whoosh again, on to the copulatory organ (in whatever form it takes).

In flies, this sperm pump evolved once but has subsequently been lost in many families, including all of the Culicomorpha (the midges and the mosquitoes) and some Bibionomorpha (the group containing the fungus gnats and the St Marks flies), along with the Sepsidae (scavenger flies), Diopsidae (stalk-eyed flies), and the Glossinidae (tsetse flies). Similar structures have been seen in Phoridae and Sphaeroceridae (lesser dung flies) as well as the good old Drosophilidae, which may turn out to be evolutionary relics of a structure these species employ, called spermatophores, which are neat parcels crammed with sperm and other fluids. These parcels are created from the secretions of the male accessory gland which enclose the sperm and other goodies, but they may not achieve a definite form till they are inside the female, an example being seen in tsetse flies.

However the sperm is transferred, it often has a long and winding journey before it reaches its target, and the process of copulation is aided by both genital and non-genital structures. Let's start with the males, as more often than not, it's the males that start it. Male genitals comprise more segments than a female's, and include both genital structures for sperm transfer, and non-genital structures, such as claspers and combs, which aid copulation by grasping on to the females. William Eberhard, an American entomologist, in 2004, compiled a comprehensive review of fly genitalia, and he found that in males nearly half of the adaptations

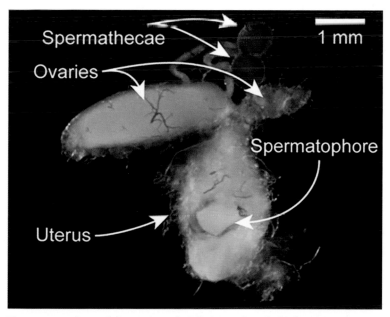

The spermatophore of the tsetse male, fully formed inside the female's uterus.

in their genitalia were with penetrative structures, to facilitate deeper sperm deposition or better positioning of the male during copulation, both of which enabled more effective sperm transfer.

The first, and sometimes the largest parts of the genitalia are the upper and lower surface of segment 9, called the epandrium (tergite 9) and hypandrium (sternite 9) respectively. In many species, the epandrium has been modified into clasping structures or forceps – the transfer of sperm may take several hours, so a tight grip is essential. Robber flies, the Asilidae, are not just amazing aerial predators, but also have some of the most conspicuous genital structures in nature. Take the genus *Efferia,* of which there are more than 240 species, described to date from Neotropical and Nearctic regions. It is equipped with what is called a helicopter tail. Neither a helicopter nor a tail, this structure is rather a pair of very grand and hairy claspers, and they are a lot larger than the rest of the abdominal segments. A brilliant example is *Efferia exaggerata*, whose name bears homage to these enormous structures. When John Smit, at the Naturalis Biodiversity Centre, described this species in 2019, he wrote about watching these individuals in flight and how they appear to drag their enormous terminalia along behind them. This must play havoc with their aerodynamic ability and this is one of the many trade-offs that males have to make. Any reduction in the flying-ability of these males must be of less disadvantage than any improvements gained by larger reproductive features.

Now this might just look like the boys are showing off, but the male has to ensure his sperm is the only sperm that produces the next generation. There is a whole lot of competition out there, not to mention things going on in the female's genital tract that forces the males to develop such exaggerated features. Males and females have been competing against each other since the sexes began, where what is good for one sex is not necessarily good for the other. The development of massive genitalia in many a fly is not just there to amuse the likes of you and me.

There are further specialized structures, including a set of paired, two-segmented regions (one gonocoxite and one gonostylus) collectively called the gonopods, and a further pair of parameres, both of which are specialized

The whole specimen and terminalia of a male *Efferia exaggerata* – an aptly named robber fly, whose greatly expanded epandrium acts as claspers.

sperm transfer appendages, which may or may not be fused with the penis.

The penis is correctly called the aedeagus, from the Greek 'aidoia' meaning genitals and 'agos' meaning leader (much better than 'penis' whose name is derived from the word for tail). It's also known as the phallus if the aedeagus has fused with the parameral sheath (a fusion of parameres). These appendages can be long, short, straight, or even twisted structures, a variety of forms resulting from the battle of the sexes.

A good example of all these bits working together is from studying *Scathophaga stercoraria*, the yellow dung fly (Scathophagidae) – a beast of a fly with some elaborate male genital structures that slot together in a very complicated puzzle. These males can often be seen hanging around dung pats and are highly aggressive to other species (they are predators), as well as to each other and to the poor females, who are often damaged as a result of their fervid attentions.

David Hosken is a researcher particularly focused on fly sexual selection and conflict, two topics that the yellow dung fly excel in. In 1999, Hosken and co-authors provided a pictorial representation of what happens when this species copulate, which in my mind resemble the Fiordlands of New Zealand. As already stated, there is not much literature associated with what actually happens during this process, due to the difficulty of making internal observations, but the Hosken team were able to prepare stained sections of the flies when viewed by a transmission electron microscope. They were able to make out the relatively long penis inserted into the female's spermathecal duct and held in place by a clasper, forceps and paramere – these males are taking no chances. They are not alone in trying to ensure that they have optimal linkage – the mosquito *Aedes aegypti* has developed lobes on the side of its penis, which form a cone shape, creating a suction cup that help keep the penis firmly lodged in place.

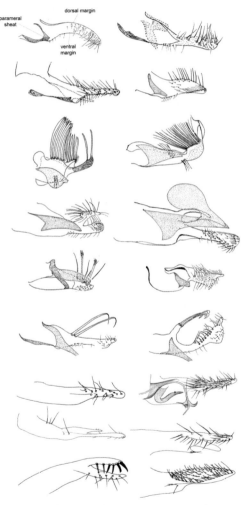

Promiscuity is not a rare event for female flies, and many have numerous mates. In fact, it's not a rare event in many animals. She mates with several males to ensure that she has a 'shopping basket' of the best produce stored inside of her, where only the best sperm gets purchased. Males have subsequently developed both mechanical as well as chemical weapons to clear the path of his enemies' sperm. Some of the favoured mechanical

The elaborate and variable parameres of different species of Phlebotominae.

mechanisms, as seen in the yellow dung fly and many other species, is to deliver his load with great force, like a projectile, to blast out any sperm already there. Male mosquitoes have a range of structures to flush out the previous occupant's sperm or persuade the female not to copulate again. *Aedes aegypti*, for example, has teeth on the end of the penis, which may shred her tract to physically prevent any further matings. But it's within *Rhamphomyia*, a commonly found genus of dance fly, that we can find perhaps the most extraordinary penis length and shape. I have used taxonomic keys that help me figure out what species are what, which describe species from this genus as having a 'tortuous' penis – I'd say it also looks a bit torturous personally.

Brutality is a feature of some fly copulations. Males with the right structures can prise open the female's genital opening, and thrust through her insides, pushing aside any body parts that may be in the way. This tactic has been found in species of tephritid fruit flies, such as the olive fruit fly *Bactrocera oleae*, where the male has a genital rod that it uses to push his sperm into the female – it is rather rough. The American dipterist Brian Brown, the phorid prince, recently sent me an image of an unknown species of *Megaselia* of that family that had an aedeagus that caused a whole group of fly experts to exclaim 'fake news', due to its incredible curly-whirly shape. Maybe he opens the wine for her as well?

The amazingly elongated penises of some male flies are not all rigid all of the time – some of them can be blown up or extended. Even I blushed at some of these structures! *Nymphomyia walkeri*, in the very primitive family of flies Nymphomyiidae, has a very long retractable aedeagus, which, when not in use, is kept neatly protected in a cone-shaped casing.

The size of a fly's copulatory appendages is not the only part of the genital setup that varies with species, their sperm size does to. The model organism, *Drosophila melanogaster*, has sperm 1.9 mm (0.075 in) long, which is large in comparison to its own body length of nearly 3 mm (0.12 in) – especially when you consider a human sperm is just 0.05 mm (0.002 in) long. *Drosophila melanogaster* are impressive, but first prize must go to its sibling *Drosophila bifurca*. This species is but a few millimetres in length (3 mm) but its sperm, once uncurled, is a whopping 5.8 cm (2¼ in) long!

This tiny animal has the longest known sperm of any animal on the planet. They are not the only *Drosophila* to have megasperm, but they are by far the most impressive. In the 1950s, Kenneth W Cooper produced his 'seminal' work on *Drosophila* sperm and wrote of its 'impressive gamete' having attained a length of 1.76 mm (0.069 in). Back then it was thought this was a giant in the sperm world – and indeed it is 300 times larger than a human sperm – but it was tiny in comparison to that of *D. bifurca*. In the wonderfully titled paper *How Long is a Giant Sperm?* published in *Nature* in 1995, Scott Pitnick and co-authors of Syracuse University, were the first to document the mega-megasperm of *Drosophila bifurca*, pointing out that it is 20 times longer than the fly that makes it, which is nearly equivalent to a human male ejaculating a sperm the length of a blue whale.

Sperm production in *Drosophila bifurca* takes 17 days to grow large enough testes for his megasperm, from about 12 mm (½ in) to 70 mm (2¾ in) that is nearly a six-fold increase in size – half of his body cavity taken over by testes. And yes, I do mean those figures. The testes of this fly are not round like humans but have been super stretched. And how does he get these megasperm into her? He basically has a pea-shooter as a sperm transferrer – pop, pop, pop go his tightly coiled sperm as he

The rather exciting corkscrew penis from a species of *Megaselia* from Costa Rica.

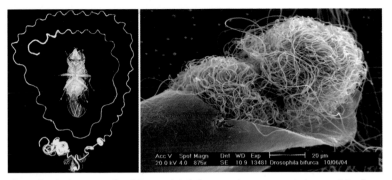

The individual sperm cells of *Drosophila bifurca*, and one untangled around the fly.

fires one spermatic pellet (containing one gamete) after another into her. She then untangles it (a very unusual event and not seen in other species with megasperm) and stores it in her own giant tubular storage organ (about 75 mm or 3 in) that is coiled inside her.

All the megasperm species may produce big sperm, but they don't produce many of them – only a few hundred during their adult life, the requirements for producing such huge gametes being so demanding. They are thought to have evolved via a Fisherian runaway process – a process of developing more and more exaggerated characteristics, a concept proposed by the British mathematical biologist Ronald Fisher. In the case of *Drosophila bifurca*, the length of the female's seminal receptacle increased over generations to prevent weak sperm from entering. As a counter to that, the male's sperm also increased in length. In studying fly sperm, we can advance our understanding of the sperm of other animals, including our own, as theirs is ethically easier to play with, as well as being larger.

Although a major component of spermatophores is sperm, there are lots of other goodies in there, too. Francesca Scolari and co-authors of the University of Pavia in 2016 looked at the contents of these parcels in tsetse flies. Along with the essential sperm, there were many proteins associated with known and unknown functions. Those known proteins play a role in facilitating the storage and maintenance of the sperm in the female and cause changes in her physiology and behaviour. And just to give you an idea of the scale of this pharmacy, male *Drosophila melanogaster* transfer more than 200 different seminal fluid proteins to the female during mating – a whole new interpretation of a protein shake. Just the presence of his ejaculate inside her starts mucking around with her behaviour; in many insects, including *Drosophila melanogaster,* the fluids increase fecundity, and also decrease her receptivity to subsequent mating – she loses her mojo. And he sure as heck does not want her fooling around with other males after all his hard work. Males even tailor their investments depending on her previous sexual activity. If a female has already been mated, the male will decrease the amount of sperm he transfers and increase the amount of proteins that will increase her fertility or decrease her libido.

Our knowledge of how mating systems work and how the individuals try to prevent promiscuity in their opposite counterparts is helpful for us in regulating the populations of what we consider pest species. The ongoing war against mosquitoes, the primary vectors for many deadly diseases in humans and our livestock, is well documented. But the mosquito's ability to evolve quickly has thwarted our attempts at controlling their populations, and insecticide resistance is widespread. The World Health Organisation (WHO) *Global Report on Insecticide Resistance in Malaria Vectors: 2010–2016* found that within the four commonly used insecticide classes – the pyrethroids, organochlorines, carbamates and organophosphates – resistance was prevalent in all the main malaria carriers across most of the globe. Designing ways to impede the success of wild populations of any species requires in-depth knowledge about how they reproduce. We may know lots about how a species behaves and how it interacts, but we often know little about its reproductive fluids. Cornell University's Ethan Degner, with co-authors were able to answer some of the pressing questions about sperm and seminal fluid from the malaria-carrying mosquito *Aedes aegypti*. In a 2018 paper, they report that there were a staggering 870 different proteins in the sperm and a further 280 in the fluids, which just goes to show you how much is happening post copulation. I find this post-mating controlling ability by males most disconcerting. But it can be both deleterious and beneficial to the female. The effect that these different seminal fluid proteins have on the females not only mucks around with her mojo, but it may also increase egg production, as well as have some antimicrobial functions, or induce her to produce her own antimicrobial peptides. This battle between the sexes results in more than just one winner – it is a fight to ensure that they both get the best offspring that genetics can buy.

Seminal fluids may also form a physical plug in the female's reproductive tract, either during or soon after mating. This is a slightly less traumatic method of blocking sperm from other males. The other way is to leave a body part behind, which may or may not result in the death of the male. It is believed that these plugs may be related to spermatophores – for instance, in the mosquito *Anopheles gambiae*, the plug was interpreted as either the left-over parts (of the case of the spermatophore) or the beginning of a new spermatophore forming. Anja Lachmann, a German dipterist, has not only spent time observing the copulatory behaviour of dung flies, but in particular has spent a fair amount of time fiddling with *Coproica vagans* plugs – all in the name of science. And in the lesser dung fly *Coproica vagans* (Sphaeroceridae), a firm but elastic mating plug is deposited by the male in the female genital opening after copulation. No subsequent suitors appear to be able to dislodge this sperm glue and it only pops out when she starts laying eggs!

Preventing other males' sperm is not the only reason why they have these plugs. They are thought to be helpful in ensuring that the owner of the plug's sperm is both protected and facilitating sperm mobility (this was shown to happen in mice). Plugs are also found in species that only mate once and the hormones present in these plugs may be one of the factors that has led to this. Many of the anopheline mosquitoes for example, are monandrous, mating once, the female then storing the sperm to use over her lifetime. Flaminia Catteruccia, at Imperial College London, and colleagues both there and abroad, are conducting some interesting work looking at the biological ingredients in these plugs, and what happens to them or because of them, after copulation, in the female. Firstly, they found that the plugs were formed by the seminal fluid proteins binding very tightly, a process controlled by a male accessory gland (MAG)-specific transglutaminase (Tgase – we use similar enzymes in food to make them more gelatinous). Catteruccia and colleagues in 2009 were able to demonstrate that, if this enzyme was not expressed, the plug didn't form, and therefore the females were not able to store any sperm.

In 2015, the same team, still playing around with these plugs, looked at some of the hormones that they contained. In *Anopheles gambiae*, the male leaves behind a plug that contains, amongst its many other constituents, the steroid hormone 20-hydroxyecdysone (20E). This hormone is quite something – it not only makes her unwilling to mate again, it also increases egg production and makes her start laying. The hormone also regulates vitellogenic lipid transporters (vitellogenin is the precursor protein of egg yolk), and so are important in egg development. But this hormone also has an unfortunate side-effect that the malarial parasites have jumped on – it weakens the female's immune system and so decreases her ability to defend herself against these invading beasties. The male's production of this hormone has resulted in a triple whammy of problems for humans – it has increased the longevity of the mozzie as the female now avoids the increased chance of predation associated with swarms; it has increased her fertility rate by boosting her egg production; and it has increased the parasite's success rates. Females naturally produce 20E but the production of this by MAGs has not been found in any other genera of mosquitoes so far or *Drosophila melanogaster*. The research into these plugs will provide us with new methods for malarial control.

Female genitalia (in flies) have historically been ignored by researchers, but this is changing thanks to our understanding of how important these parts are. Morphologically, they are challenging, but advances in microscopy are enabling us to study these structures in situ. Females store the sperm in an organ called a spermatheca, the size and number of which vary enormously. Most flies have three, but they can range in number from none to four. I picture the insides of a female fly as a chemist shop, lined with jars of different sperm that she pulls off the shelf when she is ready to concoct another life, because this is pretty much what she does. In the phlebotomine sand fly species, *Phlebotomus papatasi*, the spermathecae change both in structure and function once sperm arrives and can manipulate what sperm is released and when, to fertilize the eggs.

Those poor little sperm (or not so little with some of the *Drosophila* and so on) need further help once they are inside the female. There are secretory cells associated with the spermathecae that basically act as nanny cells, ensuring that the sperm arrive safely and then are looked after until they are needed. Sperm are quite basic, and the male's ejaculate initially provides the nutrients for their survival. But that does not last long, and this is a problem for the female as she may need this over a long time (queen bees can store sperm for decades). Some sperm really love these secretory cells – sperm from flies in the genus *Anastrepha* (Tephritidae) stick their heads into the openings of these cells. The solutes from these glands are essential to ensure that the sperm are kept healthy and mobile (they can be seen swimming around in the spermathecae, and that shows they are healthy).

Once the male has passed his sperm into the female, his role in fostering the next generation is, more often than not, over. Parental care is basically unheard of in male flies, the notable exception being *Mystacinobia zealandica*, the New Zealand bat fly. They live in communal groups in and around tree-roosting bats, and the older adult males watch over their community, including the young – they produce a 'zizzy' noise that is thought to act as a deterrent to any would-be attackers. They live in the dark and so their eyes are greatly reduced in size and so function – instead they rely a lot more on sound. Why this species lives as a group is not known, but the usual advantages of group living, such as increased protection against predators, seem likely. On top of that, this species spends a lot of time grooming each other – both larvae and adults – which means greater protection against pathogens (they live in bat poo, which can't be that pleasant).

The majority of flies lay fertilized eggs. Some species will hold onto them internally so upon laying the larvae are ready to hatch out, a term called oviviviparity. Flies don't just stop at this level of care and there are many examples of pre-birth parental care in female flies. Many species give birth to fully developed larvae, which have been nourished and protected internally up to the point of birth. Those wonderful tsetse flies are one such family, and they, along with three sibling families all found within the superfamily Hippoboscoidea (Hippoboscidae, the ked flies, and Nycteribiidae and Streblidae, the bat flies), produce just one maggot at a time. This maggot is first hatched from an egg and fed in utero (meaning in the womb) before being expelled into the world and pupating straight away, as it has no further need to feed.

Animals generally have two main strategies for reproduction: oviparity (from the Latin 'ovum' meaning egg, and 'parire' to bring forth), where eggs develop externally of the mother, and viviparity, from the Latin 'vivus' meaning living, where the mother retains the eggs and then gives birth to live young. Giving birth to live young is rarer in insects than in vertebrates, but it has still been documented in 11 insect orders including the tsetse above. Once more, the flies in general are the queens of this, with 61 examples across 22 families of flies. German dipterist Rudolf Meier and collaborators published a review in 1999 where they detailed the various types of viviparity, ranging from occasionally (she may or may not retain the larvae depending on environmental conditions) to the most advanced form, as seen in the tsetse fly.

According to Meier and co-authors, those species that can produce both eggs and larvae depending on the circumstances deserve more attention, as they are the intermediate step between oviparity and viviparity. They are often species associated with temporary habitats such as dung and rotting flesh, and as such are often found living among our livestock. Those species that can decide between laying eggs or a larva can give their young a massive advantage by laying larvae when fresh food is immediately available.

Within the viviparous species, females generally give birth to one large larva at a time (unilarviparous), but rarer are those species that give birth to more than one but less than 12 (oligolarviparous). The most typical are when large numbers of eggs or larvae develop in the uterus (multilarviparous). All the species in the acalyptrate family Ctenostylidae are multilarviparous, and OK there are only 14 species, but still. Amazingly the expectant mothers may have an eye-watering 400 larvae inside them at one time. *Ramuliseta madagascarensis* is one such species and the larvae fill not only the uterus but also both ovaries. I can't begin to imagine how uncomfortable this must be.

The majority of the tsetse flies are unilarviparous, with on average females only producing between eight and 10 offspring in her lifetime. While inside, the larvae are nurtured by what is called adenotrophic viviparity or pseudo-placental, and this is where the mother has special milk glands, the name adenotrophic means gland fed, through which she supplies milk to her offspring. In other species, larvae feed on egg yolk, feeding known as lecithotrophic.

Tsetse flies and their relatives are arguably the most extreme parental caregivers among the flies. This amazing form of nurturing was discovered by Major General Sir David Bruce, to give him his full title, an Australian born Scottish pathologist and microbiologist, and his wife Mary (who was also a microbiologist and co-author but receives less credit). Back in 1897, he found himself staying in the small village of Ubombo in northern KwaZulu, South Africa, where he had been sent to investigate a cattle disease called nagana. He found that the causative agent was a protozoan called *Trypanosoma brucei*, and that the vector that passed this among the livestock was the tsetse fly. For clarification, he did not name it after himself but rather it was named in honour of him, one of the nicest forms of scientific flattery.

Bruce was also the first to document the tsetse's unusual pregnancies. The poor mother transfers so much milk to the larvae that her lipid reserves are reduced by 50%. If there is any delay or disruption of the milk supply then birth is either delayed or, rather sadly, the maggot is aborted. The richness of the milk is crucial for the maggot's development and it contains 12 major proteins, released throughout the pregnancy. Interestingly, and of importance for biological control methods, milk is not the only thing passed from the mother. The fly has its own symbiotic bacteria – they need each other to survive – called *Wigglesworthia glossinidia*, the name of which pays homage to the great insect physiologist Sir Vincent Brian Wigglesworth, whose work I have referenced many times. These bacteria are essential for the maggot as without them the adult stage is immune-compromised and sterile. So pretty much useless. Don't panic for the poor tsetse, as it has been shown that a dietary supplement of yeast extract and vitamin B can once more restore their reproductive abilities – yep, British folks, there is once again proof that marmite is one of the best food sources on the planet. If you have not heard of this spread, you must try it.

The majority of female flies lay eggs though, and to facilitate this the terminal segments of her abdomen become thinner and often stretched or tapered, to form a very flexible egg-laying tube, the ovipositor. Imagine being able to pierce through things with your genitalia? In

A tsetse fly giving birth to one enormous larva.

many cases this is exactly what the female can do. The stretching of her abdomen is often accompanied by the thickening of the eighth segment, to help her penetrate the surface tissues of other organisms, be it vegetable, fungi or animal, and lay her precious brood. Often the ovipositor has undergone extensive hardening – the plates have become more sclerotized, which gives the mother a better tool to work with.

Among the flies with the most notable of egg-laying tubes are the Conopidae, the thick-headed flies, especially within the subfamilies Stylogastrinae and Dalmanniinae. Other common names for this family are the bee or wasp-grabbers, and this pays homage to the majority of hosts within which offspring develop – the exception is Stylogastrinae who also grab cockroaches and crickets (orthopteroids). The usual method for egg laying is to land on a host and insert the eggs, but within Stylogastrinae there are species that have developed harpoon-shaped eggs that they fire at their intended victims – a particularly girl power way of dispersing your eggs. The ovipositor of the genus *Stylogaster* is also hinged and is used to flick the eggs. We have a collection of impaled specimens in the Natural History Museum, where eggs have been catapulted in all over the body. There are records of them even piercing the eyes!

Stylogaster is the only genus in the Stylogastrinae subfamily, and among their almost 100 species are those that use marauding army ants. These ferocious and nomadic predators eat many of the fly's potential host species, especially the orthopteroids, and so to try and lay eggs on them is a dangerous tactic. The flies are pretty game to tag on to this hunting party, and many females will not survive their egg-firing activities, becoming the prey instead. Not content on harassing some of the most fearsome warriors on the planet, *Stylogaster* eggs also attach to other diptera, mostly within the calyptrates. We are not sure of all the species that are parasitized by this genus, but we are gradually getting more and more pieces to help us solve the puzzle. Marcia Couri and Gabriel Pinto da Silva Barros, both then at the National Museum in Brazil, highlight, in their 2010 paper on dipteran hosts of *Stylogaster* from Madagascar and South Africa, species found in the families Calliphoridae, Lauxaniidae, Tachinidae, Syrphidae and Muscidae.

The hosts from the Muscidae family, included genera *Stomoxys* and *Haematobosca* – those blood feeding species that are a nuisance to our livestock and many species of large mammal, including ourselves.

A group of flies where many species have developed elaborate ovipositors are the Tephritoidea superfamily. There are seven families in this group, including the economically important Tephritidae – the fruit flies, where females are all recognisable because of the shape of their tapered abdomen, often retracted within the seventh abdominal segment .

Unlike other female flies, where the eggs and faeces are released from two separate holes, in fruit flies, 'two become one', as the Spice Girls sang. This may seem most unpleasant, but it has been found in many tephritid species, including the large and important *Bactrocera* genus. Containing over 500 species, including some of the world's biggest pests the oriental fruit fly and the olive fruit fly, the mothers pass symbiotic gut bacteria at the same time as the egg to ensure a healthy offspring. The joining of the reproductive and digestive tracts forms a cloaca, a term more associated with bird morphology, which in many vertebrates is the only opening for these tracts. The word derives from the Latin for drain or sewer, and means to cleanse, in the sense of allowing waste to flow away as in a sewer. Indeed, it is a very usual feature in invertebrates, and once more highlights the amazing adaptability of flies.

Another tephritid, the apple maggot or railroad worm, *Rhagoletis pomonella,* is an economic pest on apples and other fruits, and as far back as 1890 we've known about the adult female's ability to pierce through the skin of an apple and insert her larvae into the fleshy part of the fruit. This species is unusual, not just for her hardened ovipositor, but also for mimicking a jumping spider, *Paraphidippus aurantius* (very similar to its sibling *Rhagoletis zephyria* from Chapter 6). When the fly lands, it lands face down, holding the wings up so that it indeed resembles the spider with its front legs raised in an aggressive stance – this buys it time to lay its eggs, deterring would-be predators.

In North America, apple maggots originally fed on the small fruits of the hawthorn tree, but in 1864 apple growers discovered that some individuals had switched to the larger, more succulent apples, which had been introduced to the Americas in the seventeenth century by the

The *Rhagoletis pomonella* fly deters would-be predators with its resemblance to the jumping spider *Paraphidippus aurantius*.

colonists. Today it appears that the ones that switched have formed a race that no longer feed on the hawthorn, and the ones that stayed on the hawthorn have never been found on the apples. This may constitute a case for 'sympatric speciation', i.e. where a species splits into two within the same environment but remains separate from each other.

The ovipositor on apple maggot female flies is partially retracted within the abdomen (as with all Tephritidae and Pyrgotidae – another picture-wing family), but is clearly exposed during copulation, egg-laying and scent marking. Females lace the apples with a deterrent pheromone – an oviposition-deterring pheromone (ODP) – to deter other females from laying eggs in the same apple or defecating on it.

Special chemosensilla on the ovipositors can perceive the fruit's sugars, and there are four pairs of them: one at the tip and three others in two lateral grooves, and each pair detects different chemicals. There are also mechanosensilla (hair-like and campaniform or dome-like sensilla) that respond to pressure from all directions. It is thought that these sensilla are to help determine how much force the female needs to penetrate a barrier such as vegetable skin, what the conditions are like inside the host and possibly even to monitor the transport of the eggs into the host. She is a wily lady, this one, as not only is she taking a measure of the condition of the host for her offspring, but before she has even got to this stage she is using the same sensilla to smell her male partner while *in flagrante delicto* (her intimate parts are smelling his).

This ability to read the condition of the host for her eggs is impressive, and has been determined in many other species, apart from tephritids. *Musca autumnalis*, in the Muscidae family, has mechanosensilla all over her ovipositor. In 1972, Ruth Hooper and her team at Kansas State University found that the amazing, telescoping tube of *M. autumnalis* had rings of long tactile hairs that circled the sixth and seventh abdominal segments, and that the eighth segment was also covered in tiny hairs. They discovered that, when the female penetrates into fresh cow dung, she uses these hairs to determine the depth she is at, like using a hairy dipstick.

This fly, in common with the rest of the calyptrates, was found to have, on its final, ninth abdominal segment, a pair of anal or lateral leaflets –

flaps on which the team found five different types of sensory organs, all important for sensory perception, including more hairs used for smell. It's quite something to think of your genitalia sniffing its environment.

As well as her smelling for the perfect oviposition site, some females will also release pheromones to tell others of suitable places to oviposit i.e. where to lay their eggs. Many medically important flies – mosquitoes, sand flies and black flies – release different pheromones that encourage females to oviposit at the same site. This may seem strange especially as so many species are trying to actively keep other individuals away from the best sites, but it comes down to protection, safety in numbers as it were. This was first seen with the mosquito genus *Culex* (Culicidae) where the oviposition chemical – erythro-6-acetoxy-5-hexadecanolide – attracted several species to the original ovipositing site. Large numbers of larval mosquitoes bobbing up and down to the surface prevents a build-up of scum on the water's surface, which would house pathogens detrimental to them, and more importantly the mother increases her chance of having more of her offspring survive if they are dispersed in a much larger group. Having all of offspring together also improves their chances of finding a mate as the adults all emerge at the same place.

The structures of both male and female fly terminalia are some of the most complicated and convoluted features to be found in any of the orders. It's exciting to think what we might learn about the function of so many as-yet unexplored structures. The variability across species will keep us busy for a long time, as only a few species have been studied in any great detail. But creatures that can develop sperm twenty times their body size, or provide over a thousand different proteins in their ejaculate, are definitely worth a closer look.

The end

End? No, the journey doesn't end here.
Death is just another path. One that we all must take.

J.R.R. Tolkien, The Return of the King

I AM WRITING this final chapter while at fly school. Not a school for flies, but an international course that is run by the Los Angeles Museum of Natural History in what I have discovered to be a rather foggy part of California. There are 25 students from across the globe attending, from various fly disciplines and at various points in their career. The tutors are some of the greatest dipterists around, all of which I have featured in this book, thanks to their efforts in the subject. The students have flies that they have caught in the Californian countryside in front of them, and they are trying to figure out what family/genus/species of fly they are looking at. And there may be some mutterings about 'costal breaks' and further morphological headaches. But there is also a sense of wonder at the individuals that they are staring at down a microscope. Fat flies, fluffy flies, long flies, little flies – so much information bundled up into small, moveable packages. We have been studying flies for many years, inside and out, and hopefully, we've learned something about them along the way. But even so, we have basically only scratched the cuticle: there are colossal amounts still to learn about this group, not least how many flies are related to each other, as well as what many of the different parts of the body do.

The head of *Ceratitis capitata* with its semaphore seta.

Flies, it would appear, are able to fool us simple humans regularly with their adaptability, with many red herrings to disrupt our understanding, – for instance, the many occasions on which two species look alike but are in fact not closely related at all. Take the family Tachiniscidae, close relatives of Tephritidae (fruit flies), which David Grimaldi, of the American Museum of Natural History, has just been telling us about at one of the lectures. When you look at the species found in this family, you begin to understand some of the challenges that still face us. These little beasts at first glance look just like Tachinidae (the parasite flies) – they are both very bristly, which may be common in calyptrates, but it is not in acalyptrates. These similarities do not help us understand their relationship with each other – which is not close at all – but it may help us understand something about the overlap in their lifestyles and the functionality of their morphology; for example, both Tachiniscidae and Tachinidae have parasitic larvae. Superficially, there are many families, genera and species of fly that resemble each other, all of which are worthy of investigation.

Camposella insignata with its enormous paddle-shaped antennae.

Then there are things that look so odd we just can't work out why. *Camposella insignata*, for instance, one of the hunched-back flies in the family Acroceridae, has an astonishing pair of antennae that look like two table-tennis bats. It was originally described in 1919 by Frank Cole, an American dipterist, who amusingly wrote that 'it was small wonder that the entomologists at the National Museum thought that they were dreaming when they came across it'.

Why they have such marvellous antennal paddles is unknown. It is only seen in the males, so we can but presume it is a secondary sexual feature. But this fly, as with many flies, is hard enough to find, let alone to study its courtship. But why should this happen in this species and not closely related ones? It's a question that hangs.

There are some lovely expanded antennae in the genus *Porpocera*, (Stratiomyidae, soldier flies), with just two species, found only in South Africa and Zimbabwe. Both species have enlarged antennae, with the first five flagellomeres greatly engorged and merged together. This can be found in both sexes, so is therefore not a sexual feature – so what is going on here? When Martin Hauser and co-authors wrote about *Porpocera* in the *Manual of Afrotropical Diptera* in 2017, nothing was known about the biology of this species, and we're still largely in the

A *Porpocera horrida* female with its greatly enlarged antennae.

dark. We can sadly add it to that rather long list of amazing-looking creatures we know nothing about.

But while we're trying to work it all out, we can have some fun. When we give names to species, we often make reference to something in their morphology, some kind of distinguishing feature. There are other examples of nomenclature, for example using people and places, but a morphological trait is a good one. Some have gone further and started naming species after the physical attributes of people well-known to them. There's a little horse fly, for instance, which was discovered in 2011 by the Australian researcher Bryan Lessard. A dense patch of golden hairs at the end of its abdomen makes it look as if it has a pair of gold hot pants on, and in honour of this attribute, Lessard named the species *Scaptia beyonceae*, after, of course, Beyoncé herself.

There is a code that taxonomists are supposed to follow when it comes to the naming of species, a set of guidelines that are continuously revised by the International Commission of Zoological Nomenclature (ICZN). We have already seen in this book examples of species that have been named in honour of researchers. But there is no rule against naming species after celebrities. American researcher Brian Brown discovered the wingless female of a new species of phorid in 2018. This species had more than likely been overlooked before, as it measures a diminutive 0.395 mm in length, and as such is the smallest species of fly found so far. Its species name, though, comes from an entirely different physical characteristic. And so, Brown called his new fly *Megapropodiphora arnoldi*, because: 'as soon as I saw those bulging legs, I knew I had to name this one after Arnold Schwarzenegger'. Although it has one pair of bulging legs, the other two appear to be much reduced – it obviously needs to mix up its gym routine.

Many species of flies stand out for their beauty rather than their brawn, with amazingly patterned or metallic exoskeletons. And some have both. Take the gorgeous *Amenia imperialis* (of the calliphorids), commonly called the snail-killing fly. The males of this species have a wonderfully conspicuous yellow head to accompany the standard metallic bodies that we see in the more commonly encountered species, as well as some very obvious white spots. The name 'snail-killing fly' is earned by the fact it is thought that the large internally reared larvae are parasites of land snails. Once more, although this is a common species in and around forests and gardens in eastern Queensland and New South Wales in Australia, we understand little about its life history. But if a bright yellow head does not scream 'COME AND STUDY ME', then I don't know what does!

Flies have not only been ahead of the game, but in many ways have changed the game. Just when you thought that flies couldn't get more adaptable, you come across a species that blows your mind. Another Australian fly – that country has produced many odd creatures – is the very rare *Badisis ambulans*, a strange-looking species in the stilt-legged family, the Micropezidae. There are approximately 500 species of these long-legged flies, mostly winged. But this one is rather different. Both the larval and the adult stage have adapted to their environments in very different ways. First off, the larvae are found in the Albany pitcher plant (*Cephalotus follicularis*) of Western Australia, where they can tolerate the plant's digestive enzymes and live in their deep flowers without fear of being dissolved. Not only that, they move around using their

The snail-killing blow fly – *Amenia imperialis*.

spines and suckers called 'creeping welts', ripping apart and feeding upon the less well-adapted insects that have fallen into these miniature death chambers. These little brutes metamorphose into strange-looking adults; adults that have no wings or halteres, and there is a constriction between the abdominal segments one and two that look like very small waists, all of which makes them look incredibly ant-like. Why look like a defenceless fly when you can look like a highly aggressive and powerful ant? This species of fly is found in the same habitat as the *Iridomyrmex* conifer species-group of ants whom they mimic well. The larvae and the adult habitat are also quite different: the adults must walk a fair distance to their preferred forest type. There have been so many adaptations in this one species to enable it to live how it does, and we have yet to look closely at the behavioural or chemical traits of the adult – does it wave its legs around as other hymenopteran mimics do, or copy the hydrocarbons of the ants to evade being attacked by them?

A *Badisis ambulans* – the ant-mimic whose larvae feast in the Albany pitcher plant.

Living in pitcher plants does ostensibly appear to be a risky tactic, but more species of diptera have been identified living in them than of any other organism to date – 169 species of flies, which include mosquitoes, midges, crane flies, hover flies and house flies. The adaptation to this aquatic environment that is toxic to most other creatures is something to be admired.

But if you think that is impressive, then what about a species of fly that scuba dives in alkaline lakes? *Ephydra hians*, commonly called the alkali fly, belongs to the family Ephydridae – the brine flies – and is found across North America. They are found in abundance around the salty waters of Mono Lake in California, where the tenacious mothers are able to submerge themselves in the alkaline waters and lay their eggs at the bottom. The ability to do this has been known for a long time, and the novelist Mark Twain even wrote of them back in 1872, that, 'You can hold them under water as long as you please – they do not mind it; they are only proud of it'. How exactly they do this is only just being understood. As the female enters the water, tiny hairs on her

The scuba-diving fly *Ephydra hians*.

thorax and elsewhere trap air, enabling an air bubble to form around her, and transforming her into a natural submersible. These smart little things are not the only animals to be able to scuba dive – lots of types of beetles form these 'plastrons', or air bubbles. But it is this fly's unique ability to, survive in such a salt environments, that elevates this species above all others.

Floris van Breugel and Michael H Dickinson, both at the California Institute of Technology, published a paper in 2017 that explained the process in more detail. By attaching flies to a rod, Breugel and Dickinson determined that the females used a force roughly 18 times their body weight to push head first through the water's surface. Their cuticles, as with all flies, are covered with tiny hairs and layers of waxy hydrocarbons. These hairs are hydrophobic (they repel water) and so, as the fly crawls into the water the hairs trap a thin layer of air that forms a bubble around them, acting like an external lung. *Ephydra hians* can descend to depths of 8 m (26¼ ft) for up to 15 minutes. But this is a very saline environment that would normally reduce the ability of the fly to stay dry. But not in alkali flies, thanks to the unique composition of their hydrocarbon layer. It is non-polar – meaning it has no electrical charge – and that acts as a barrier over the cuticle below, which, because it has a weak charge, would normally pull ions towards it, and attract water. Is there nothing that these hydrocarbons can't do?

Other flies have adapted to survive in other extremes too – some in the most bizarre fashion. Many species of fly will allow themselves to fall into a coma to protect themselves during times of extreme temperature. This is because flies are 'poikilotherms' (from the Greek 'poikilos', meaning varied) – a term used to describe animals whose internal temperature can vary significantly – they cannot maintain a constant body temperature in the same fashion that you or I can. But flies are able to allow themselves to be supercooled: they can take their bodies below freezing point without all the internal cells becoming solid. This allows them to survive during periods of extreme cold, by inducing a state called a chill-coma. Some species that can instigate chill-comas have glycerol in their cells that acts as antifreeze. One such fly is *Chymomyza costata*, on which Tomáš Štětina and fellow authors, all in the Czech Republic published a paper in 2018. And guess what? It's another species of drosophilid.

They subjected the larvae of this species to a series of lower and lower temperatures – supercooling to -10°C (14 °F), freezing at -30°C (-22° F), and cryopreservation at -196°C (-321°F). And some survived... to be fair, there was quite a high death rate in the individuals at -196°C – but still, some survived, which means that these flies are the most complex animal to be able to do so. These temperatures are of course extreme, and not often encountered by many animals, but in understanding how these flies cope with these conditions, we can learn to develop techniques for helping humans cope with adverse conditions.

Whilst undertaking research for this book, I felt a growing unease that I could never produce a popular science book that gives sufficient credit to the animals that I love, or the research that is being undertaken on them. The best I can hope for is that you see it as a mere flyby of all things Diptera. It is impossible to do justice to such a charismatic group of animals. We have been studying their morphology for hundreds of years and have made massive leaps in understanding their genetics over the past hundred. They adapt incredibly well. They wriggle themselves into countless terrestrial niches, as well as some marine, and have developed some amazing structures to help them do so.

As I sit on my sofa and stare up at that house fly, flying around my light, I think right now that fly is processing 250 images a second, smelling and listening with an incredibly complex and highly developed matrix of

Chymomyza costata larva that was frozen to -32°C and then -196°C.

receptors, linked by a neurological pathway that we have only just begun to understand, whilst reacting to stimuli in ridiculously short timeframes, and all the time monitoring its environment for predators, food and most importantly sex. And all of this is going on in a body that is shorter than my thumbnail. What fantastic beasts we share our lives with.

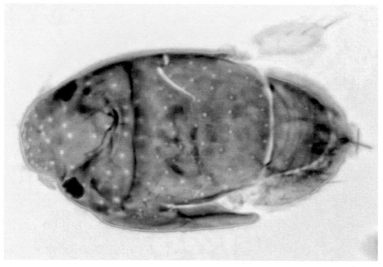

Megapropodiphora arnoldi – the tiniest of species that packs the mightiest of punches (in dipterological glee that is).

INTRODUCTION

Bhadra, P. et al. (2014), Factors affecting accessibility to blowflies of bodies disposed in suitcases. *Forensic Science International*, 239: 62–72.

Butler, M.G. (1982), A 7-year life cycle for two *Chironomus* species in Arctic Alaskan tundra ponds (Diptera: Chironomidae). *Canadian Journal of Zoology*, 60 (1): 58–70

Connor, S. (2006), *Fly*. Reaktion Books, Chicago, USA.

Hebert, P.D.N. et al. (2016), Counting animal species with DNA barcodes: Canadian insects. *Philosophical Transactions of the Royal Society B.* 371 (1702).

Larsen, B. et al. (2017), Inordinate fondness multiplied and redistributed: the number of species on Earth and the new pie of life. *The Quarterly Review of Biology*, 92 (3): 229–265.

Mohr, S. (2018), *First in Fly: Drosophila Research and Biological Discovery*. Harvard University Press, USA.

Pliny, the Elder (1469), *Libros Naturalis Historiæ Nouitiu Camenis Qritiu Tuo Opus Natu Apud Me Proxima Fetura Licentore Epistola Narrare Consului Tibi Iocundissimi Imperator*. Johann Speyer, Venice.

Townson, H. et al. (2013), Systematics of *Anopheles barbirostris* van der Wulp and a sibling species of the *Barbirostris* Complex (Diptera: Culicidae) in Eastern Java, Indonesia. *Systematic Entomology*, 38 (1): 180–191.

CHAPTER 1 PRE-ADULTHOOD

Arcifa, M.S. (2000), Feeding habits of Chaoboridae larvae in a tropical Brazilian reservoir. *Brazilian Journal of Biology*, 60 (4): 591–597.

Banerjee, A. et al. (2018), Two species of *Caiusa* Surcouf (Diptera: Calliphoridae) new to India, with data on larval behaviour and morphology. *Biodiversity Data Journal*, 6: e27736.

Briones, M.D.L. et al. (2013), Identification of human remains by DNA analysis of the gastrointestinal contents of fly larvae. *Journal of Forensic Sciences*, 58 (1): 248–250.

Courtney, G.W. et al. (2019), Biodiversity of diptera. In: *Insect Biodiversity: Science and Society*, Foottit, R.G. and Adler, P.H. (eds.). Blackwell Publishing, Oxford.

Crosskey, R.W. (1990), *The Natural History of Blackflies*. Natural History Museum, London.

Darwin, C. (1839), *Voyages of the Adventure and Beagle, Volume III. Journal and Remarks. 1832–1836.* Henry Colburn, London.

Hall, M.J.R. et al. (2017), The 'dance' of life: visualizing metamorphosis during pupation in the blow fly *Calliphora vicina* by X-ray video imaging and micro-computed tomography. *Royal Society Open Science*, 4: 160699.

Hayes, M.J. et al. (2016), Identification of nanopillars on the cuticle of the aquatic larvae of the drone fly (Diptera: Syrphidae). *Journal of Insect Science*, 16 (1): 36; 1–7.

Ichiki, R. and Shima, H. (2003), Immature life of *Compsilura concinnata* (Meigen) (Diptera: Tachinidae). *Annals of the Entomological Society of America*, 96 (2): 161–167.

Ingram, B.A. (2011), Population dynamics of chironomid larvae (Diptera: Chironomidae) in earthen fish ponds in south-eastern Australia. *Asian Fisheries Science*, 24 (1): 31–48.

Jackman R. et al. (1983), Predatory capture of toads by fly larvae. *Science*, 222 (4623): 515–516.

Järbrink, K. et al. (2017), The humanistic and economic burden of chronic wounds: a protocol for a systematic review. *Systematic Reviews*, 6: 15.

Kennedy, H.D. (1958), Biology and life history of a new species of mountain midge, *Deuterophlebia nielsoni*, from eastern California (Diptera: Deuterophlebiidae). *Transactions of the American Microscopical Society*, 77 (2): 201–228.

Lenhard, R.E. (1973), *William Stevenson Baer: A Monograph*. Privately printed.

Lindegaard, C. and Jónasson, P.M. (1979), Abundance, population dynamics and production of zoobenthos in Lake Mývatn, Iceland. *Oikos*, 32 (1/2): 202–227.

MacArthur, R.H. and Wilson, E.D. (1967), *The Theory of Island Biogeography*. Princeton University Press, Princeton.

Maitland, D.P. (1992), Locomotion by jumping in the Mediterranean fruit-fly larva *Ceratitis capitata*. *Nature*, 355: 159–161.

Mathison, B.A. and Pritt, B.S. (2014), Laboratory identification of arthropod ectoparasites. *Clinical Microbiology Reviews*, 27 (1): 48–67.

McAlister, E. (2017), *The Secret Life of Flies*. Natural History Museum, London.

Roberts, M.J. (1971), On the locomotion of cyclorrhaphan maggots (Diptera). *Journal of Natural History*, 5 (5): 583–590.

Shen, T. et al. (2017), Remotely triggered locomotion of hydrogel mag-bots in confined spaces. *Scientific Reports*, 7 (1): 16178.

Swammerdam, J. (1669), *Historia Insectorum Generalis*. Utrecht.

Tripathy, A. et al. (2017), Natural and bioinspired nanostructured bactericidal surfaces. *Advances in Colloid and Interface Science*, 248: 85–104.

Truman, J.W. and Riddiford, L.M. (1999), The origins of insect metamorphosis. *Nature*, 401 (6752): 447–452.

van der Plas, M.J.A. et al. (2009), Maggot secretions suppress pro-inflammatory responses of human monocytes through elevation of cyclic AMP. *Diabetologia*, 52 (9): 1962–1970.

Walshe, B.M. (1947), The function of haemoglobin In *Tanytarsus* (Chironomidae). *Journal of Experimental. Biology*, 24: 343–351.

Wotton, R.S. and Hirabayashi, K. (1999) Midge larvae (Diptera: Chironomidae) as engineers in slow sand filter beds. *Water Research*, 33. 1509–1515.

CHAPTER 2 HEADS UP

Andersen, T. et al. (2016), Blind Flight? A new Troglobiotic orthoclad (Diptera, Chironomidae) from the Lukina Jama – Trojama Cave in Croatia. *PLOS ONE*, 11 (4): e0152884.

Caro, T. et al. (2019), Benefits of zebra stripes: behaviour of tabanid flies around zebras and horses. *PLOS ONE*, 14 (2): e0210831.

Collett, T.S. and Land, M.F.J. (1975), Visual spatial memory in a hoverfly. *Comparative Physiology*, 100 (1): 59–84.

Dodson, G.N. (1989), The horny antics of antlered flies. *Australian Natural History*, 12: 604–611.

Franceschini, N (2014), Small brains, smart machines: from fly vision to robot vision and back again. *Proceedings of the IEEE*, 102 (5): 751–775.

Gilbert, C. (1994), Form and function of stemmata in larvae of holometabolous insects. *Annual Review of Entomology*, 39: 323–349.

Helfrich-Förster, C. et al. (2002), The extraretinal eyelet of *Drosophila*: development, ultrastructure, and putative circadian function. *Journal of Neuroscience*, 22 (21): 9255–9266.

Hofbauer, A. and Buchner, E. (1989), Does *Drosophila* have seven eyes? *Naturwissenschaften*, 76: 335–336.

Joern, A. and Rudd, N.T. (1982), Impact of predation by the robber fly *Proctacanthus milbertii* (Diptera: Asilidae) on grasshopper (Orthoptera: Acrididae) populations. *Oecologia*, 55 (1): 42–46.

Kral, K. and Meinertzhagen, I. (1989), Anatomical plasticity of synapses in the lamina of the optic lobe of the fly. *Philosophical Transactions of The Royal Society B Biological Sciences*, 323 (1214): 155–83.

Marshall, S. (2012), *Flies: The Natural History and Diversity of Diptera*. Firefly Books Ltd, Canada.

Martín-Vega, D. et al. (2010), Back from the dead: *Thyreophora cynophila* (Panzer, 1798) (Diptera: Piophilidae) 'globally extinct' fugitive in Spain. *Systematic Entomology*, 35 (4): 607–613.

Moorefield, H.H. and Fraenkel, G. (1954), The character and ultimate fate of the larval salivary secretion of *Phormia regina* Meig. (Diptera, Calliphoridae). *Biological Bulletin*, 106 (2): 178–184.

Nguyen, T.C. (2003), A new species of *Diathoneura* (Diptera: Drosophilidae) from Costa Rica with a striking sexual dimorphism. *Kansas Entomological Society*, 76: 104–108.

Petersen, M.J. et al. (2010), Phylogenetic synthesis of morphological and molecular data reveals new insights into the higher-level classification of Tipuloidea (Diptera). *Systematic Entomology*, 35 (3): 526–545.

Seifert, P. et al. (1987), Internal extraocular photoreceptors in a dipteran insect. *Tissue and Cell*, 19 (1): 111–118.

Swammerdam, J. et al. (1776), *Bybel der Nature*.

Taylor, G.M. and Krapp, H.G. (2007), Sensory systems and flight stability: what do insects measure and why? *Advances in Insect Physiology*, 34: 231–316.

Wardill, T.J. et al. (2017), A novel interception strategy in a miniature robber fly with extreme visual acuity. *Current Biology*, 27 (6): 854–859.

Wendt, L.D. and Ale-Rocha, R. (2015), Antlered richardiid flies: new species of *Richardia* (Tephritoidea: Richardiidae) with antler-like genal processes. *Entomological Science*, 18: 153–166.

Xiang, Y. et al. (2010), Light-avoidance-mediating photoreceptors tile the *Drosophila* larval body wall. *Nature*, 468 (7326): 921–926.

Zheng, Z. et al. (2018), A complete electron microscopy volume of the brain of adult *Drosophila melanogaster*. *Cell*, 174: 730–743.

CHAPTER 3 THE ANTENNAE

Borkent, A. and Belton, P. (2006), Attraction of female *Uranotaenia lowii* (Diptera: Culicidae) to frog calls in Costa Rica. *Canadian Entomologist*, 138 (1): 91–94.

Carson, R. (1962), *Silent Spring*. Houghton Mifflin Company, Boston.

Chen, L. and Fadamiro, H.Y. (2007), Behavioral and electroantennogram responses of phorid fly *Pseudacteon tricuspis* (Diptera: Phoridae) to red imported fire ant *Solenopsis invicta* odor and trail pheromone. *Insect Behavior*, 20: 267–287.

Clements, A.N. (1956), The antennal pulsating organs of mosquitoes and other Diptera. *Journal of Cell Science*, 3 (97): 429–433.

Jackson, J.C. and Robert, D. (2006), Nonlinear auditory mechanism enhances female sounds for male mosquitoes. *Proceedings of the National Academy of Sciences of the United States of America*, 103 (45): 16734–16739.

Liu, Z. et al. (2016), Identification of male- and female-specific olfaction genes in antennae of the oriental fruit fly (*Bactrocera dorsalis*). *PLOS ONE*, 11 (2): e014778.

Sivinski, J. et al. (1984), Acoustic courtship signals in the Caribbean fruit fly, *Anastrepha suspensa* (Loew). *Animal Behaviour*, 32 (4): 1011–1016.

van Breugel, F. et al. (2015), Mosquitoes use vision to associate odor plumes with thermal targets. *Current Biology*, 25 (16): 2123–2129.

Wang, Y. et al. (2016), Morphology, ultrastructure and possible functions of antennal sensilla of *Sitodiplosis mosellana* Géhin (Diptera: Cecidomyiidae). *Journal of Insect Science*, 16 (1): 93; 1–12.

Yeates, D.K. and Wiegmann, B.M. (1999), Congruence and controversy: toward a higher-level phylogeny of Diptera. *Annual Review of Entomology*, 44: 397–428.

Zhang, D. et al. (2013), Sensory organs of the antenna of two Fannia species (Diptera: Fanniidae). *Parasitology Research*, 112 (6): 2177–2185.

Zhang, D. et al. (2016), The antenna of horse stomach bot flies: morphology and phylogenetic implications (Oestridae, Gasterophilinae: *Gasterophilus* Leach). *Scientific Reports*, 6: 34409.

Zhanga, Z. (2016), Morphology, distribution and abundance of antennal sensilla of the oyster mushroom fly, *Coboldia fuscipes* (Meigen) (Diptera: Scatopsidae). Systematics, morphology and biogeography. *Revista Brasileira de Entomologia*, 60 (1): 8–14.

CHAPTER 4 MOUTHPARTS

Barraclough, D. and Slotow, R. (2010), The South African keystone pollinator *Moegistorhynchus longirostris* (Wiedemann, 1819) (Diptera: Nemestrinidae): notes on biology, biogeography and proboscis length variation. *African Invertebrates*, 51 (2): 397–403.

Catts, E.P. and Garcia, R. (1963), Drinking by adult *Cephenemyia* (Diptera: Oestridae). *Annals of the Entomological Society of America*, 56 (5): 660–663.

Darwin, C. (1862), *Fertilisation of Orchids*. John Murray, London.

Drukewitz, S.H. et al. (2019), Toxins from scratch? Diverse, multimodal gene origins in the predatory robber fly *Dasypogon diadema* indicate a dynamic venom evolution in dipteran insects. *GigaScience*, 8: 1–13.

Gurera, D. et al. (2018), Lessons from mosquitoes' painless piercing. *Journal of the Mechanical Behavior of Biomedical Materials*, 84: 178–187.

Jung, J.W. et al. (2015), A novel olfactory pathway is essential for fast and efficient blood-feeding in mosquitoes. *Scientific Reports*, 5: 13444.

Karolyi, F. et al. (2012), Adaptations for nectar-feeding in the mouthparts of long-proboscid flies (Nemestrinidae: Prosoeca). *Biological Journal of the Linnean Society*, 107 (2): 414–424.

Khramov, A.V. and Lukashevich, E.D. (2019), A Jurassic dipteran pollinator with an extremely long proboscis. *Gondwana Research*, 71: 210–215.

Lehnert, M.S. (2017), Mouthpart conduit sizes of fluid-feeding insects determine the ability to feed from pores. *Proceedings of the Royal Society B*, 284 (1846): 20162026.

Maeda, T. et al. (2014), Neuronal projections and putative interaction of multimodal inputs in the subesophageal ganglion in the blow

fly, *Phormia regina. Chemical Senses*,
39: 391–401.

Martín-Vega, D. et al. (2018), Micro-
computed tomography visualization of
the vestigial alimentary canal in adult
oestrid flies. *Medical and Veterinary
Entomology*, 32 (3): 378–382.

Whitfield, F.G.S. (1925), The relation
between the feeding-habits and the
structure of the month-parts in the
Asilidae (Diptera). *Proceedings of the
Zoological Society of London*, 95 (2):
599–638.

Wizen, G. (2018), Little transformers:
Forcipomyia, the midge that turns
into a balloon. http://gilwizen.com/
forcipomyia/

CHAPTER 5 THE THORAX

Arthur, B.J. and Hoy, R.R. (2006),
The ability of the parasitoid fly *Ormia
ochracea* to distinguish sounds in the
vertical plane. *Journal of the Acoustical
Society of America*, 120 (3): 1546–1549.

Frey, R. (1941), Die gattungen und
arten der dipterenfamilie Celyphidae.
Notulae Entomologicae, 21: 3–17.

Navarro, J.C. et al. (2010), Highest
mosquito records (Diptera: Culicidae)
in Venezuela. *Revista de Biologia
Tropical*, 58 (1): 245–254.

Stubbs, A. and Drake, M. (2001),
*British Soldierflies and Their Allies:
A Field Guide to the Larger British
Brachycera*. British Entomological &
Natural History Society.

CHAPTER 6 THE WINGS

Andersen, T. et al. (2016), Blind flight?
A new troglobiotic orthoclad (Diptera,
Chironomidae) from the Lukina Jama
– Trojama cave in Croatia. *PLOS ONE*,
11 (4): e0152884.

Colyer, C.N. and Hammond, C.O.
(1951), *Flies of the British Isles*.
Frederick Warne, London.

Daltorio, K.A. and Fox, J.L. (2018),
Haltere removal alters responses to
gravity in standing flies. *Journal of
Experimental Biology*, 221: 181719.

Graciolli, G. and Dick, C.W. (2012),
Description of a second species of
Joblingia Dybas & Wenzel, 1947
(Diptera: Streblidae). *Systematic
Parasitology*, 81 (3): 187–193.

Hoffmann, J. et al. (2018), A simple
developmental model recapitulates
complex insect wing venation patterns.
*Proceedings of the National Academy of
Sciences*, 115 (40): 9905–9910.

Hooke, R. (1665), *Micrographia, or
some physiological descriptions of minute
bodies made by magnifying glasses, with*

observations and inquiries thereupon. J.
Martyn and J. Allestry, London.

Jobling, B. (1939), On the African
Streblidae (Diptera Acalypterae)
including the morphology of the genus
Ascodipteron Adens and a description
of a new species. *Parasitology*, 31
(2): 147.

McAlpine, J.F. et al. (eds.) (1981),
Manual of Nearctic Diptera. Volume 1.
Agriculture, Canada.

Monica, H. et al. (1987), A sheep in
wolf's clothing: tephritid flies mimic
spider predators. *Science*, 236 (4799):
308–310.

Mou, X. and Sun, M. (2012), Dynamic
flight stability of a model hoverfly in
inclined-stroke-plane hovering. *Journal
of Bionic Engineering*, 9: 294–303.

Mountcastle, A.M. and Combes, S.A.
(2014), Biomechanical strategies
for mitigating collision damage in
insect wings: structural design versus
embedded elastic materials. *Journal
of Experimental Biology*, 217 (7):
1108–1115.

Nilssen, A.C. and Anderson, J.R.
(1995), Flight capacity of the reindeer
warble fly *Hypoderma tarandi* (L.) and
the reindeer nose bot fly *Cephenemyia
trompe* (Modéer) (Diptera: Oestridae).
Canadian Journal of Zoology, 73:
1228–1238.

Pringle, J.W.S. (1948), The gyroscopic
mechanism of the halteres of Diptera.
*Philosophical Transactions of the Royal
Society B: Biological Sciences*, 233:
347–384.

Prokop, J. et al. (2017), Paleozoic
nymphal wing pads support dual
model of insect wing origins. *Current
Biology*, 27: 263–269.

Raad, H. et al. (2016), Functional
gustatory role of chemoreceptors in
Drosophila wings. *Cell Reports*, 15 (7):
1442–1454.

Sotavalta, O. (1953), Recordings
of high wing-stroke and thoracic
vibration frequency in some midges.
Biological Bulletin, 104: 439–444.

Wiesenborn, W.D. (2011), UV-excited
fluorescence on riparian insects except
Hymenoptera is associated with
nitrogen content. *Psyche*, 2011: 875250.

CHAPTER 7 LEGS

Brues, C.T. (1900), Peculiar tracheal
dilatations in *Bittacomorpha clavipes*
Fabr. *Biological Bulletin*, 1 (3):
155–160.

Cook, R.M. (1977), Behavioral role
of the sex combs in *Drosophila
melanogaster* and *Drosophila*

simulans. Behaviour Genetics, 7
(5): 349–357.

Daugeron, C. et al. (2010), Extreme
male leg polymorphic asymmetry in
a new empidine dance fly (Diptera:
Empididae). *Biology Letters*, 7 (1):
11–14.

Downes, J.A. (1978), Feeding
and mating in the insectivorous
Ceratopogoninae (Diptera). *Memoirs
of the Entomological Society of Canada*,
110 (S104): 1–62.

Evenhuis, N.L. (2013), The
Campsicnemus popeye species group
(Diptera: Dolichopodidae) from
French Polynesia. *Zootaxa*, 3694:
271–279.

Gorb, S.N. and Heepe, L. (2018),
Biological fibrillar adhesives:
functional principles and biomimetic
applications. *Handbook of Adhesion
Technology*, 1641–1676.

Grimaldi, D. and Underwood, B.A.
(1986), *Megabraula*, a new genus
for two new species of Braulidae
(Diptera), and a discussion of braulid
evolution. *Systematic Entomology*, 11
(4), 427–438.

Kanmiya, K. (1990), Acoustic
properties and geographic variation
in the European courtship signals
of the European chloropid fly,
Lipara lucens meigen (Diptera,
Chloropidae). *Journal of Ethology*, 8
(2): 105–120.

Martin, S.J. and Bayfield, J. (2014), Is
the bee louse *Braula coeca* (Diptera)
using chemical camouflage to
survive within honeybee colonies?
Chemoecology, 24 (4): 165–169.

Minakawa, N. et al. (2007), Predatory
capacity of a shorefly, *Ochthera
chalybescens*, on malaria vectors.
Malaria Journal, 6: 104.

Richards, O.W. (1927), Sexual
selection and allied problems in the
insects. *Biological Reviews*, 2 (4):
298–364.

Samoh, A. et al. (2017), Eight
new species of marine dolichopodid
flies of Thinophilus Wahlberg, 1844
(Diptera: Dolichopodidae) from
peninsular Thailand. *European Journal
of Taxonomy*, 329 :1–40.

Simpson, K.W. (1975), Biology and
immature stages of three species
of Nearctic Ochthera (Diptera:
Ephydridae). *Proceedings of the
Entomological Society of Washington*,
77 (1): 129–155.

Soler C. et al. (2004), Coordinated
development of muscles and tendons
of the *Drosophila* leg. *Development*,
131 (24): 6041–6051.

Thoma, V. et al. (2016), Functional. dissociation in sweet taste receptor neurons between and within taste organs of *Drosophila*. *Nature Communications*, 7: 10678.

West, T. (1862), The foot of the fly; its structure and action: elucidated by comparison with the feet of other insects. *Transactions of the Linnean Society of London*, 23: 393–421.

Young, J.H. and Merritt, D.J. (2003), The ultrastructure and function of the silk-producing basitarsus in the Hilarini (Diptera: Empididae). *Arthropod Structure & Development*, 32 (2–3): 157–65.

CHAPTER 8 THE ABDOMEN

Brown, B.V. and Porras, W. (2015), Extravagant female sexual display in a *Megaselia Rondani* species (Diptera: Phoridae). *Biodiversity Data Journal*, 2015; (3): e4368.

Cerretti, P. et al. (2014), First report of exocrine epithelial glands in oestroid flies: the tachinid sexual patches (Diptera: Oestroidea: Tachinidae). *Acta Zoologica*, 96 (3): 383–397.

de O Gaio, A. et al. (2011), Contribution of midgut bacteria to blood digestion and egg production in *Aedes aegypti* (Diptera: Culicidae) (L.). *Parasites & Vectors*, 4: 105.

Green, L.F.B. (1979), The fine structure of the light organ of the New Zealand glow-worm *Arachnocampa luminosa* (Diptera: Mycetophilidae). *Tissue and Cell*, 11 (3): 457–465.

Greene, C.T. (1924), Synopsis of the North American flies of the genus *Scellus*. *Proceedings of the United States National Museum*, 65 (16): 1–18.

Miguel-Aliaga, I. et al. (2018), Anatomy and physiology of the digestive tract of *Drosophila melanogaster*. *GENETICS*, 210 (2): 357–396.

Pont, A. (1987), 'The mysterious swarms of sepsid flies': an enigma solved? *Journal of Natural History*, 21 (2): 305–317.

Su, K. et al. (2017), Sex ticklers and dirty flies: The development and evolution of a novel abdominal appendage in male sepsid flies. *Mechanisms of Development*, 145: S21.

CHAPTER 9 THE TERMINALIA

Barták, M. and Kubík, Š. (2015), Palaearctic Species of *Rhamphomyia* (*Pararhamphomyia*) *anfractuosa* group (Diptera, Empididae). *Zookeys*, 514: 111–127.

Couri, M. and da Silva Barros, G.P.

(2009), Diptera hosts of *Stylogaster* Macquart (Diptera, Conopidae) from Madagascar and South Africa. *Revista Brasileira de Entomologia*, 54 (3): 361–366.

Galati, E.A.B. et al. (2017), An illustrated guide for characters and terminology used in descriptions of Phlebotominae (Diptera, Psychodidae). *Parasite*, 24: 26.

Hooper, R.L. et al. (1972), Sense organs on the ovipositor of the face fly, *Musca autumnalis*. *Annals of the Entomological Society of America*, 65 (3): 577–586.

Hosken, D.J. et al. (2011), Internal female reproductive anatomy and genital interactions during copula in the yellow dung fly, *Scathophaga stercoraria* (Diptera: Scathophagidae). *Canadian Journal of Zoology*, 77 (12): 1975–1983.

Ilango, K. (2005), Structure and function of the spermathecal complex in the phlebotomine sandfly *Phlebotomus papatasi Scopoli* (Diptera: Psychodidae): II. Post-copulatory histophysiological changes during the gonotrophic cycle. *Journal of Biosciences*, 30 (5): 733–747.

Inatomi, M. et al. (2019), Proper direction of male genitalia is prerequisite for copulation in *Drosophila*, implying cooperative evolution between genitalia rotation and mating behavior. *Scientific Reports*, 9 (1): 210.

Li, Y. et al. (2013), A new species of *Ocydromia meigen* from China, with a key to species from the Palaearctic and Oriental. regions (Diptera, Empidoidea, Ocydromiinae). *ZooKeys*, 349 (349): 1–9.

Mattei, A.L. et al. (2015), Integrated 3D view of postmating responses by the *Drosophila melanogaster* female reproductive tract, obtained by micro-computed tomography scanning. *Proceedings of the National Academy of Sciences of the United States of America*, 27.

Mitchell, S.N. et al. (2015), Evolution of sexual traits influencing vectorial capacity in anopheline mosquitoes. *Science*, 347: 985–8.

Pitnick, S. et al. (1995), How long is a giant sperm? *Nature*, 375 (6527): 109.

Rogers, D.W. et al. (2009), Transglutaminase-mediated semen coagulation controls sperm storage in the malaria mosquito. *PLOS Biology*, 7 (12): e1000272.

Samoh, A. et al. (2017), Eight new species of marine dolichopodid flies of

Thinophilus Wahlberg, 1844 (Diptera: Dolichopodidae) from peninsular Thailand. *European Journal of Taxonomy*, 329: 1–40.

Scolari, F. et al. (2016), The spermatophore in *Glossina morsitans morsitans*: insights into male contributions to reproduction. *Scientific Reports*, 6: 20334.

Smit, J.T. (2019), Robber flies from Sint Eustatius, Lesser Antilles, with the descriptions of *Efferia exaggerata* sp. n. and the male of *Ommatius prolongatus* Scarbrough (Diptera: Asilidae). *Zootaxa*, 4586 (1): 141–150.

THE END

Brown, B. (2018), A second contender for "world's smallest fly" (Diptera: Phoridae). *Biodiversity Data Journal*, 6: e22396.

Cole, F.R. (1919), A new genus in the dipterous family Cyrtidae from South America. *Entomological News and Proceedings of the Entomological Section of the Academy of Natural Sciences of Philadelphia*, 30 (10), 270–274.

Hauser, M. et al. (2017), Stratiomyidae (soldier flies). In: *Manual of Afrotropical Diptera, vol. 2, chapter 41: Nematocerous Diptera and lower Brachycera*. Pretoria, SANBI Graphics & Editing.

Lessard, B. and Yeates, D. (2011), New species of the Australian horse fly subgenus *Scaptia* (*Plinthina*) Walker 1850 (Diptera: Tabanidae), with species descriptions and a revised key. *Australian Journal of Entomology*, 50 (3): 241–252.

McAlpine, D.K. (1990). A new apterous micropezid fly (Diptera: Schizophora) from Western Australia. *Systematic Entomology*, 15 (1): 81–86.

van Breugela, F. and Dickinsona, M.H. (2017), Superhydrophobic diving flies (*Ephydra hians*) and the hypersaline waters of Mono Lake. *Proceedings of the National Academy of Sciences*, 114 (51): 13483–13488.

PICTURE CREDITS

p.2, 59 ©US Geological Survey/Science Photo Library; p.6 ©Wellcome Collection; p.12,67, 87, 230 ©Solvin Zankl/naturepl.com; p.18 ©Brendan B. Larsen et al., *Inordinate Fondness Multiplied and Redistributed: the Number of Species on Earth and the New Pie of Life,* The Quarterly Review of Biology 92, no. 3 (September 2017): 229-265; p.22 ©S.Rae/Flickr; p.24 ©Fabrice Parais; p.26, 37, 264 ©Kim Taylor/naturepl.com; p.27 ©Dom Greves; p.29 (top), ©Matt Bertone; (middle) ©Filippo Bortolon; (bottom) ©Hakon Haraldseide; p.31 ©Richardi, Vinicius Sobrinho et al, *Morpho-histological characterization of immature of the bioindicator midge Chironomus sancticaroli Strixino and Strixino (Diptera, Chironomidae),* Revista Brasileira de Entomologia, Volume 59, Issue 3, July–September 2015, P 240-250; p.32 ©meiningi/Shutterstock.com; p.35 ©Blaine A. Mathison, Bobbi S. Pritt, *Laboratory Identification of Arthropod Ectoparasites,* Clinical Microbiology Reviews, 27(1):48-67, January 2014; p.38 ©Eye Of Science/Science Photo Library; p.42 ©Kyle Schnepp; p.51 ©Sanford Porter/USDA; p.52 ©Hall Martin J. R. et al., *The 'dance' of life: visualizing metamorphosis during pupation in the blow fly Calliphora vicina by X-ray video imaging and micro-computed tomography* 4R. Soc. open sci.; p.54, 225 ©Martin Dohrn/naturepl.com; p.56 (middle) ©Claude Nuridsany & Marie Perennou/Science Photo Library; (bottom) ©Warwick Sloss/naturepl.com; p.58 ©Clouds Hill Imaging Ltd/Science Photo Library; p.61 (top) ©Pascal Guay/Shutterstock; (bottom) ©Z. Zheng et al, *A Complete Electron Microscopy Volume of the Brain of Adult Drosophila melanogaster,* Cell, Volume 174, Issue 3, 26 July 2018, Pages 730-743. e22; p.64 ©Buschbeck, E.K., Friedrich, M. *Evolution of Insect Eyes: Tales of Ancient Heritage, Deconstruction, Reconstruction, Remodeling, and Recycling.* Evo Edu Outreach 1, 448–462 (2008); p.68 ©Katz B and Minke B (2009). Drosophila photoreceptors and signaling mechanisms. Front. Cell. Neurosci. 3:2. doi: 10.3389/neuro.03.002.2009; p.71 ©David Maitland/naturepl com; p.72 ©Thomas Shahan/Science Photo Library; p.74 ©Steven Russell Smith Ohia/Shutterstock; p.78 ©Paul Bertner/Minden/naturepl.com; p.81 ©Paul Bertner; p.88 ©Judy Gallagher/Flickr; p.89 ©Z.Zhang, *Morphology, distribution and abundance of antennal sensilla of the oyster mushroom fly, Coboldia fuscipes (Meigen) (Diptera: Scatopsidae),* Rev. Bras. entomol. vol.60 no.1 São Paulo Jan./Mar. 2016; p.90 ©Zhang, D. et al., The antenna of horse stomach bot flies: morphology and phylogenetic implications (Oestridae, Gasterophilinae: Gasterophilus Leach). Sci Rep 6, 34409 (2016); p.95 ©Avitabile D. et al., Mathematical modelling of the active hearing process in mosquitoes7J. R. Soc. Interface; p.96 ©Graham Wise from Brisbane, Australia / CC BY (https://creativecommons.org/licenses/by/2.0); p.103 ©Itamar Ofer; p.105 ©Chris Raper; p.108 ©Kelsey Bailey; p.111 ©Scott Bauer/US Department of Agriculture/Science Photo Library; p.114 ©Ozgur Kerem Bulur/Shutterstock; p.117 Redrawn from http://web.csulb.edu/~dlunderw/entomology/18-Mecop_siphon_Diptera.pdf; p.118 ©Alexander V.Khramov, Elena D. Lukashevich, *A Jurassic dipteran pollinator with an extremely long proboscis,* Gondwana Research, Volume 71, July 2019, Pages 210-215; p.119 ©Ramon M Batlle; p.123, 209 ©Steve Gschmeissner/Science Photo Library; p.126 ©Torsten Dikow; P.127, 128 ©Gil Wizen; p.130 ©Stephan Holger Drukewitz et al, *A Dipteran's Novel Sucker Punch: Evolution of Arthropod Atypical Venom with a Neurotoxic Component in Robber Flies (Asilidae, Diptera),* Toxins 2018, 10(1), 29; p.132 ©Lehnert Matthew S. et al., *Mouthpart conduit sizes of fluid-feeding insects determine the ability to feed from pores* 284 Proc. R. Soc. B; p.133 ©Mancomunidad/Flickr; p.135 (top), 205 ©Power and Syred/Science Photo

ACKNOWLEDGEMENTS

To all in Diptera – thank you – your work is inspirational. And to the Natural History Museum, London another thank you, to the Publishing staff who forced me into this and then kept me on a tight leash – I couldn't (and wouldn't) have done this without you – and everyone in Life Sciences that helped me, especially the Dipteran Team who have carried and shouldered my work load when I have not been there. I particularly want to thank Nigel Wyatt, whose knowledge, assistance and humour was invaluable, and I will forever be grateful, and the Dipterists Forum and all the other organisations that I have leaned on for information. Thank you to the world of social media – my endless requests for references, images, help and guidance did not fall on deaf ears and I owe so much of the information included in this book to all that helped me – you are all stars.

I would like to thank my friends for moral support, especially Matt, Ruth, Rusc, Nicky, Polly, Jack (and all NatSCA Chums), Paul (and all your family), and the many more that have been amazing. A large shout out to Miranda, Beulah and Hannah – you girls have been my rock. Can I please give a massive special thanks to Team Dominica – thanks for the support, giggles and rum that helped with my creative flare. And thank you too to my family – all of it – including cousins, godparents, halfsiblings and the rest of this goofy assemblage – I don't know that I deserved to be born into this medley, but I am grateful.

A special thank you to my reviewers Matthew Cobb and Peter Chandler – sorry for all the torture that I put you through. And another special thank you as well to Dave Hall and Phil de Montmorency for being my civilian testers and editors. Dave – once again – I could not have done this without you – you are a truly skilled editor, a fellow adventurer and a real friend. Phil – thank you for all you have brought and all you have dealt with, and may you continue to do so – you are amazing. But most of all I thank Alfie. My Monster of Darkness and provider of headbutts.